"An authoritative yet readable account by a distinguished medical scientist. This book will likely change your lifestyle and lead to popular demand for environmental changes."

—Norman F. Cantor, author of the *New York Times* best-seller *In the Wake of the Plague*

"Tierno takes the reader on a masterful journey. He serves up intricate concepts in a delicious, palatable manner that only a lifetime of battlefield experience can bestow. Extremely informative and useful."

—Edward J. Bottone, Ph.D., professor of medicine, microbiology, and pathology, Mount Sinai School of Medicine

"In this expertly written book, Tierno shares his encyclopedic knowledge of microorganisms along with lessons from evolution, archaeology, history, immunology, and his research to create a comprehensive reference."

—Melanie Maslow, M.D., F.A.C.P., chief, Infectious Diseases Section, New York Veterans Administration Medical Center, and assistant professor of clinical medicine, New York University School of Medicine

"Authoritative and highly readable. . . . Fascinating tales of scientific detection . . . interesting insights . . . and practical advice."

—Bruce S. McEwen, Ph.D., Alfred E. Mirsky Professor and head of Harold and Margaret Milliken Hatch Laboratory of Neuroendocrinology, The Rockefeller University

Also by Philip M. Tierno, Jr., Ph.D.

The Secret Life of Germs:
Observations and Lessons from a Microbe Hunter

Available from POCKET BOOKS

PROTECT YOURSELF AGAINST
BIOTERRORISM

Philip M. Tierno, Jr., Ph.D.

POCKET BOOKS
NEW YORK   LONDON
TORONTO SYDNEY SINGAPORE

The ideas, procedures, and suggestions in this book are intended to supplement, not replace, the medical advice of trained professionals. All matters regarding your health require medical supervision. Consult your physician before adopting the medical suggestions in this book, as well as about any condition that may require diagnosis or medical attention. The authors and publisher disclaim any liability arising directly or indirectly from the use of this book.

An *Original* Publication of POCKET BOOKS

 POCKET BOOKS, a division of Simon & Schuster, Inc.
1230 Avenue of the Americas, New York, NY 10020

Copyright © 2002 by Philip M. Tierno, Ph.D.

ISBN: 0-7434-5350-6

First Pocket Books printing January 2002

10 9 8 7 6 5 4 3 2 1

POCKET and colophon are registered trademarks of Simon & Schuster, Inc.

For information regarding special discounts for bulk purchases, please contact Simon & Schuster Special Sales at 1-800-456-6798 or business@simonandschuster.com

Printed in the U.S.A.

To all the victims of the
World Trade Center disaster
and
the post–September 11, 2001, bioterrorism attacks.
They paid the ultimate price
for freedom.

ACKNOWLEDGMENTS

I wish to thank the staff at Pocket Books, particularly senior editors Tracy Bernstein and Mitchell Ivers, for their enthusiasm and guidance in seeing this book to fruition. Many thanks to my literary agent, Robert Tabian, for his continuing to "take care of matters" for me. I am most appreciative for the contributions made by my secretary Ann Sullivan, and my typist Tina Sullivan.

I could not have completed the task of writing this book without the cooperation and understanding of my wonderful family. Thanks to my wife, Josephine, my daughters, Alexandra and Meredith, and their husbands, Francois and Thomas. Their unfailing support provided the necessary impetus to complete my endeavor.

"The only thing we have to fear . . .
. . . is fear itself!"
—FRANKLIN DELANO ROOSEVELT, 1941

CONTENTS

PROTECT
YOURSELF
AGAINST
BIOTERRORISM

INTRODUCTION

We are at a unique place in America's history. Simply put, we have been savagely violated. We hurt deeply. We shall never forget. In fact, America will never be the same again. For 225 years the face of terror showed itself only in a very limited way. The carnage of September 11th changed all that. In its aftermath a more ominous specter reared its ugly head— bioterror! The subject had long been talked about by leading governmental and health agencies, always evoking mixed opinions with regard to our state of awareness and our preparedness to deal with germ warfare. Ready or not, our preparedness was tested in the fall of 2001. While we did not fail that test miserably, it is clear that we were not as prepared as we should have been. Especially because we had been forewarned ten years earlier, in August 1991, when representatives of the Iraqi government announced to a United Nations Special Commission Team that they had conducted research into offensive use of agents of bioterror such as *Bacillus anthracis* and botulinum toxins. According to a U.S. Army research report, dated January 1991, Iraq deployed one hundred R400 bombs with botulinum toxin, fifty with anthrax, and two with aflatoxin during the Persian Gulf War. We knew then that it was only a matter of time before some bioterroristic event would befall us.

Until October 2001, there had never been any deaths from bioterrorism reported in the United States. The only documented case in the U.S. had been the 1984 incident in which members of the Rayneeshi cult poisoned salad bars with *Salmonella* in Dallas, Oregon. Although no one was killed in that incident, eight hundred people were made sick and fatalities could have easily occurred among the elderly and people with weakened immune systems. Certainly the 1993 bombing of the World Trade Center, the 1995 bombing of the Murrah Federal Building in Oklahoma City, and the subsequent horrendous complete destruction of the World Trade Center buildings in 2001 had already conclusively shown that the United States is not immune to foreign or homegrown terrorists.

On September 11, 2001, following the unimaginable terrorist attacks on New York City and Washington, D.C., the Centers for Disease Control (CDC) recommended that the nation increase its surveillance for any unusual disease occurrences, or clusters of disease, asserting that these may be sentinel indicators of a bioterrorist (BT) attack. Eerily, as predicted, cases of anthrax were reported in Florida, then in New York City and Washington, D.C. New Jersey soon followed, thereby justifying the CDC's suspicions. Indeed a bioterrorist was at large. The media frenzy that accompanied these anthrax reports was unprecedented in American history. Public fears and anxiety gripped the country and initially were further fueled by the apparent misinformation provided by both governmental sources and the so-called medical/scientific experts. The public received so much contradictory information that it was both confused and angered. Understandably, our learning curve was a little longer than we had imagined.

Certainly over the years one thing about germ warfare has been made crystal clear: Not only is it attractive to governments, it is equally attractive to terrorist cells, organizations, and even disgruntled individuals for the same reasons. It delivers the biggest bang for the buck, and it is relatively easy to carry out. It also comes with a big bonus—it has a dramatic psychological effect on the population!

Compared with nuclear weapons or conventional armaments on ships, tanks, and planes, biological weapons are relatively cheap and easy to make. For any individual or organization with a rudimentary understanding of microbiology and the requisite materials, making biological agents of death in quantity is little more demanding technically than brewing beer. The microorganisms involved are often readily obtainable in nature, like the anthrax bacilli that abound in soils in the Middle East, or can easily be acquired from sources such as a country's pharmaceutical and agricultural industries, as well as from many prestigious academic institutions. Toxic ricin, for example, which strikes at the central nervous system, can be extracted from the same castor beans that are the source of castor oil.

The United States Department of Defense has published a list of the seventeen likeliest biological weapons. They fall into three categories. The first includes deadly bacteria such as anthrax and plague germs. The second comprises viruses, such as those which cause smallpox, encephalitis, and hemorrhagic fevers like Ebola, Lassa, and Rift Valley fever. The last group is made up of toxins that attack the central nervous system, such as botulinum, fungal toxins, and ricin. There is one other category of weapon not included in the original list of seventeen, which I've added for the sake of completeness—chemical agents.

POTENTIAL BIOLOGICAL AGENTS	VACCINE AVAILABLE
Anthrax	Yes
Botulinum toxins	Yes
Brucellosis	Yes
Cholera	Yes
Clostridium perfringens toxins	No
Crimean Congo hemorrhagic fever	Yes
Melioidosis	No
Plague	Yes
Q fever	Yes
Ricin	Yes (experimental)
Rift Valley fever	Yes
Saxitoxins	No
Smallpox	Yes
Staphylococcal enterotoxin B, TSST-1	Yes (experimental)
Trichothecene mycotoxins	No
Tularemia	Yes
Venezuelan equine encephalitis	Yes
Chemical agents and gases	No

Vaccines are available for many of these biological agents, as you can see above. But in the absence of vaccination programs for the general public, the non-experimental vaccines have so far been used in the United States only to protect soldiers going overseas, counterterrorism units, and some so-called first responders—the emergency medical, fire, and police personnel who would be dispatched to the site of a bioterrorism attack to try to contain the damage and treat the victims.

Delivering the right treatment to victims would be one of the most troublesome aspects of any act of bioterrorism. The first indication of an attack would likely come some time after the fact, as people began falling ill in an unusual way, as was evidenced by the anthrax cases that cropped up post-9/11! Unusual patterns of illness take time to emerge. The incubation period that usually intervenes between exposure to an infectious germ and the first appearance of symptoms in infected people will be one source of delay. Another will be the problem of recognizing whether an illness has a natural cause or not, especially if its early stages mimic familiar disorders like the flu. These two factors will handicap medical personnel even if they have immediate access to effective antibiotics and antitoxins. By the time a correct diagnosis is made, it may well be too late to save lives.

As far as terrorists are concerned, the time lag

between exposure to a biological agent and the appearance of symptoms may be both a plus and a minus. On the one hand, an incubation period of a few days or longer gives terrorists time to attack and escape before an alarm is sounded. On the other hand, too long an incubation may diffuse the impact that the terrorists are hoping to achieve in order to bring attention to themselves, their cause, and their grievances.

In this light we can see that anthrax deserves its number-one ranking on the top eighteen list because of a combination of circumstances. It is highly lethal, relatively fast-acting, and it can be delivered efficiently in aerosol form by sprayer or crop duster, subject of course to the vagaries of weather and terrain. Or, as we have already witnessed, a series of anthrax-laced letters can effectively raise sufficient fear in the population even though this delivery mechanism doesn't usually carry a high casualty rate. Other biological agents, like the nerve gas sarin, can also be delivered in droplet nuclei that can injure or kill a person if they are inhaled or even if they are brought into contact with bare skin. But they are likely to be more effective in an enclosed space than in open air, limiting the number of potential victims. This does not mean that the results of such an attack would not be devastating. When the radical Japanese religious sect Aum Shinriyko released sarin nerve gas in the Tokyo subway system in 1995, scores of people were killed or injured. The possible spread of germ-warfare agents in the New York City subway system has been tested with *Serratia marcesens,* a relatively innocuous germ that grows as red-colored colonies, which allows its presence to be detected easily. It was found that the suction of trains traveling through the New York City subway tunnels could easily carry germs from one station to another.

Germs that an impatient terrorist might reject as too slow-acting might prove formidable weapons in a war. A highly contagious germ like *Yersinia pestis,* the cause of pneumonic plague, which can be passed from one person to another in sweat and respiratory secretions, could deliver a vicious one-two punch, as people made sick by an act of war infected others before they died. Even the most sophisticated health-care system would surely buckle under the strain of such a modern plague, if one were ever unleashed.

WHY THIS BOOK

In an attempt to provide the public with some important basic information regarding the agents of bioterrorism—a frame of reference, if you will—I have put together this compendium. I hope that a public forearmed with reliable information will be in a better position to understand and interpret the wealth of data being presented to it via the media. In addition, I hope that such an information base will help calm the fear and anxiety that has already developed over the recent anthrax attacks on the United States. The assumption here is that information could be the best means of reducing or mollifying public fear.

Some may question the need for the degree of detail provided by this book, but I believe that the public will be best served by having as complete a resource as possible at their disposal to help them make everyday decisions for their particular circumstance in life.

In summary, this manual provides:

- Available information on each biologic agent listed, including pertinent background and history.

- Clinical features of illnesses caused by each agent, including incubation period, signs and symptoms, diagnosis, treatment, prevention, and prophylaxis. This section also details pre- and post-exposure management if available.

- Some recommendations for citizens in the form of **Protective Response Strategies** in order to minimize their risk of contracting bioterror agents, including discussion of significant decontamination information with readily available household products.

A SHORT HISTORY OF
BIOTERRORISM: PRE-9/11

Biological warfare began in the distant past, when the first enemy waterhole was poisoned. For a long time this was probably the limit of biowarfare. New biological weapons or tactics depended on advances in knowledge, or lucky guesses, about how infectious diseases spread. The first recorded instance of more extensive biowarfare occurred in A.D. 1346, when Tatars besieging Italian traders in a citadel in Crimea threw corpses of plague victims over the wall, starting a vogue in siege tactics that lasted as long as the Black Death itself did.

Developments in biological warfare and bioterrorism have inevitably kept step with developments in medical science generally. In this respect Pasteur,

Koch, and Ehrlich opened Pandora's box when they made the discoveries that gave birth to modern microbiology. With the promise of miracle germ cures came the threat of horrific new germ killers.

In World War I, chemical warfare with poison gases was the other side of the magic bullets of Dr. Ehrlich's chemotherapy. The casualties of these weapons totaled 800,000 men on both sides. In World War II, all the major powers conducted germ-warfare research, and some countries went further. Josef Mengele injected Jews and other prisoners at Auschwitz with typhoid and tuberculosis germs. The Japanese formed a large unit for biowarfare in 1936. Based in Japanese-occupied Manchuria, in northern China, the noble-sounding Epidemic Prevention and Water Supply Unit, or Unit 731, experimented on captured Chinese, Soviet, and American soldiers. During the late 1930s and early 1940s, the Japanese dropped "disease" bombs on Chinese cities with an unknown loss of life.

The United States did not deploy such weapons, but it had them. It manufactured anthrax and botulinum bombs in the expectation that they might be needed to retaliate against Germany's or Japan's use of biological agents. At the end of the war, conscious that biological weapons could well become more important, the United States shielded Japan's Unit 731 from prosecution for war crimes in exchange for its research data and expertise, just as Hitler's V-2 rocket scientists, such as Wernher Von Braun, were brought into America's missile program.

After World War II, a biological-arms race proceeded in parallel with the nuclear-arms race, except that biological-weapons capability spread around the world farther and faster because of its relatively low cost and low technical demands. To take two recent examples of this ongoing trend, in 1988 Libya built a chemical-weapons plant in the guise of a pharmaceutical factory. And Iraq used mustard gas in its long war with Iran, and has used both mustard gas and toxic nerve agents against its own dissident Kurdish population. Saddam Hussein's stockpile of weapons of mass destruction is thought to include many other germ-warfare agents, as well.

Throughout recent decades, efforts have been made to limit or ban the use of biological weapons. The 1972 Convention on the Prohibition of the Development, Production, and Stockpiling of Biologic and Toxic Weapons and Their Destruction has been signed by one hundred countries, including the United States. But if Saddam Hussein's example is any indication, it has been honored at least as much in the breach as in the observance. Often this occurs with the cooperation of companies in the West that sell so-called dual-use technology to countries like Iraq and Libya. A plant for making pesticides can readily be turned into one for making biological weapons, for example. In the United States, a 1996 law, the Antiterrorism and Effective Death Penalty Act, increases penalties for development of biological weapons or misuse of germs to spread disease. Under this law, authorized germ-culture repositories, like the American Type and Culture Collection in Washington, D.C., must follow strict protocols to verify that the use of germs is legitimate before shipping any cultures. Likewise, some scientists have suggested that it be made illegal under international law to participate in any biological- or chemical-weapons activity that violates international treaties.

The pace of biological-weapons development may well be accelerating, thanks to innovations in genetic and molecular engineering. In 1997, *The Cobra Event*, a novel by Richard Preston, revolved around the release of a genetically engineered mixture of smallpox and common cold virus. Reading it inspired President Bill Clinton to push for increased funding to study and defend against bioterrorism. In an eerie example of life imitating art, a team of Australian scientists announced in January 2001 that they had unintentionally created a killer mousepox virus. Working with a weakened virus that usually makes mice only slightly ill, the researchers wanted to create a mouse contraceptive that would trigger an immune response to the mouse's own eggs. They genetically altered one part of the mousepox virus by inserting a gene that directs the production in interleukin-4, an important immune-system chemical in human beings as well as mice. Instead of acting as a contraceptive, however, the result wipes out a mouse's entire immune system. The mousepox virus cannot hurt people, but the implication of this research is that any virus that does afflict human beings, like one of the common cold viruses, could be turned into a killer virus in the same way.

Or perhaps the smallpox virus will be made even deadlier than in its natural form. Smallpox, which is estimated to have killed half a billion people throughout history, more than all wars and other epidemics combined, has been eradicated in nature. This achievement stands as one of the triumphs of modern medicine. The last recorded case occurred in Africa in 1977, and most countries stopped vaccinating against the disease in 1980. As the capstone of medicine's victory over smallpox, practically all stocks of smallpox viruses and vaccines were supposed to have been destroyed by April 1999. Shortly before that date President Clinton ordered that a collection of smallpox viruses and vaccines should be maintained for security purposes until June 2002. The reason for this was the finding by United States intelligence agencies that other collections were being maintained by Russia, Iraq, Iran, and North Korea, among other countries. If another country or a terrorist group unleashed smallpox on the United States, perhaps in a genetically altered form, we would desperately need a collection of virus strains and vaccines that we could work with to find ways to limit the damage. Speaking to this issue, Nobel laureate Dr. Joshua Lederberg has said, "We have no idea what may have been retained, maliciously or inadvertently, in the laboratories of a hundred countries from the time that smallpox was a common disease. These would be the likely sources of supply for possible bioterrorists."

Evidence to support Dr. Lederberg's fear came to the attention of the world media in June 2000. In Vladivostok, Russia, that month, groups of six-to-twelve-year-old children found and played with ampoules of weakened smallpox that had been negligently discarded by an old vaccine collection. Because the smallpox was in a weakened form from which vaccines could have been made, the children survived infection, albeit at the cost of permanently scarred faces. It is interesting to note that the doctors who treated the children did not at first know what to do. Since smallpox had been eradicated in nature more than twenty years before, none of them could recognize the telltale signs of humanity's age-old enemy.

The relative cheapness and ready availability of biological weapons, the potential for genetically engineering designer plagues, all these factors have led the FBI and other law-enforcement agencies to suspect that it is a question of when, not if, bioterrorism takes lives in the United States. Threats to spread anthrax have become fairly frequent recently. Although so far they have turned out to be hoaxes, sooner or later we seem vulnerable to being hit with the real McCoy. In 1998, for example, the FBI arrested two would-be terrorists in Las Vegas, Nevada. The two men said they planned to release anthrax bacilli into the atmosphere, and they did indeed have anthrax, but it proved to be a nonpathogenic vaccine strain. In January 2001, twenty-five or more similar threats were received across Canada and the United States. A typical case occurred at a Wal-Mart in Victoria, British Columbia, where a letter was received that claimed to contain anthrax. Fortunately it did not, but before that could be conclusively established the clerk who opened the letter was given a precautionary dose of ciprofloxacin.

AGROTERRORISM

Instead of trying to spread disease directly in a population, terrorists might attack indirectly by poisoning a country's food supply. Agroterrorism, as it is coming to be called, is emerging as an increasing thrust. U.S. Department of Agriculture administrator Floyd Horn has said that "a biological attack [on America's crops or livestock] is quite plausible."

Many diseases that take a devastating toll on food animals, such as foot-and-mouth disease (FMD), do not affect human beings at all. Spreading such diseases would be technically easy and a low-risk strategy for terrorists, who would need to take no special precautions in handling and deploying disease-causing germs. But there would be enormous costs to the victim country. The 2001 outbreak of FMD in Great Britain will wind up costing several billion dollars at minimum, in addition to being profoundly demoralizing for the entire country. The epidemic started on a single farm. It spread with devastating quickness, because the virus that causes FMD can easily be transmitted in a number of ways, including by airborne particles and even by people's shoes tracking dirt from infected areas to other places. Although this outbreak happened naturally and is not linked to terrorism, it suggests the extent of the damage that terrorists could inflict on an entire country, or a large region of a country, simply by infecting the herds on one or a few farms.

Imagine then that a terrorist group or criminal organization wreaked havoc on the agricultural industry in a country or a region of a country. To bring attention to a cause or to extract money, the perpetrators could threaten to do the same thing again in a sort of biological protection racket. This seemed like an outlandish plot when Ian Fleming used it in the mid-1960s in his James Bond novel *On Her Majesty's Secret Service*. Today it seems more and more a genuine possibility. Alternatively, organized criminals might use agroterror to manipulate commodities markets. Advance knowledge that certain crops or herds are going to be tainted could be a means to insure huge profits from trading in futures contracts for other crops and herds.

In this connection it is important to note that the spread of FMD in Great Britain in 2001 was helped along by government cuts in inspection and food-safety measures. Just as the United States became complacent about tuberculosis, and slashed funds for surveillance and treatment only to have multi-drug-resistant tuberculosis (MRTB) threaten to assume epidemic proportions, so Great Britain slashed budgets for surveillance of its food and livestock. In 1991 Great Britain had 43 regional animal health offices and 330 government veterinarians. Today it has half as many offices and many fewer vets, as well as fewer government labs for disease testing, even though the threat of food-borne illness and livestock disease has been increasing year by year. It is sobering to realize that the virus that causes FMD was the first animal virus to be discovered by science, in 1898. A full century later, we still can do little more to halt an epidemic of this terrible disease than to quarantine infected animals and destroy them, often in the hundreds of thousands.

To avoid such disasters, whether from natural or unnatural causes, the United States and other countries must remain aggressively on guard against

FMD and other easily transmittable animal and crop diseases. In the United States, existing safeguards may not be enough to protect against the threats that, in a globally connected world, could arrive at any time at an airport near you. In April 2001, the U.S. Department of Agriculture and the Federal Emergency Management Agency conducted an agroterrorism war game, using computer-generated models, that also involved officials from the departments of Defense, Commerce, Interior, Energy, and Health and Human Services. The object of the game was to figure out how to respond to an outbreak of FMD in a farm state like Iowa. According to the *New York Times*'s account of the exercise, it showed that the outbreak would quickly spread to other states and that stopping it would require "the combined strength of all federal disaster agencies, including the military."

In light of these threats and evidence of biological-weapons development in so many countries around the world, including a number that are bitterly opposed to the United States, our society urgently needs improved abilities to detect, prevent, and limit bioterrorism and to treat its victims. In recognition of this, between 1998 and 2001 Congress more than doubled funding for efforts to combat biological, chemical, nuclear, and radiological terrorism, from $645 million to $1.5 billion per year. The country's best **Protective Response Strategy** will surely evolve dramatically as a result of these expenditures. Among other elements, an effective national **Protective Response Strategy** might benefit from the following recommendations:

- A coordinated plan is needed that networks all federal, state, and municipal antibioterrorism programs. All first responders must be adequately educated and equipped to deal with any biowarfare or bioterrorism event.

- Military and law-enforcement personnel and medical and health practitioners, especially first responders, should be vaccinated whenever possible so that they can more effectively carry out their respective missions.

- Research and development of new-wave vaccines and antibiotics needed to prevent and treat disease caused by biological weapons should be stepped up.

- Vaccination programs for the general public should be instituted to protect against the likeliest biological weapons, such as anthrax.

- The public should be kept informed about germ warfare and should be instructed to report unusual activities to local authorities. "Bioterrorism watches" could be introduced on the model of neighborhood crime watches.

- All physicians and health-care providers should familiarize themselves with the symptoms and treatment of the diseases caused by the eighteen likeliest potential biological-warfare agents.

- A nationwide epidemiological surveillance program should link all medical facilities with the CDC or another assigned federal agency in order to identify clusters of cases that might have occurred. Small clusters may signal a terrorist practicing before a larger-scale act is carried out.

The keynote to all this is vigilance. Any complacency or overconfidence will surely prove fatal, sooner or later. The bottom line is that we must remain on the alert for animal and human epidemics and keep searching for better ways to respond to them, whether they are of natural or unnatural origin.

1.
ANTHRAX

Bioterror Agent	*Bacillus anthracis*
Type of Weapon	Bacterium
Disease Caused	Anthrax
Transmissible (person-to-person)	No (except via secretions from cutaneous lesions)
Incubation Period	1 to 60 days
Cardinal Features of Disease	*Inhalation:* flu-like without nasal discharge, then acute respiratory distress with mediastinal lymph widening *Cutaneous:* blister to ulcer with black eschar (scab) *Gastrointestinal*: GI distress with ulceration, vomiting of blood
Treatment	Antibiotics: ciprofloxacin, doxycycline, others
Vaccines Available	Yes

HISTORY AND BACKGROUND

Bacillus anthracis, the causative agent of anthrax, is really a zoonotic disease—that is, a disease that usually infects animals—but humans are also very susceptible. *B. anthracis* is a rod-shaped bacterium that produces a resting phase called a spore. It stains blue with aniline dyes; thus it is said to be gram-positive. (Aniline dyes are used to stain bacteria in order to aid in their identification. Bacteria that stain blue are said to be gram-positive, those that stain red are said to be gram-negative.) Ordinarily when bacteria grow they are said to be in a vegetative form, but when conditions are not ideal for growth, certain bacteria are able to sporulate, or produce spores (which are resting forms of bacteria). Thus, they can survive even under adverse conditions. Such is the case for *B. anthracis.* The vegetative forms of these bacteria are usually much easier to kill with simple germicides, such as alcohol or peroxide, or even with heat, but spores are very resistant to both chemicals and heat. In fact, *B. anthracis* spores can live for decades while remaining fully infectious. This is important because the infective form of *B. anthracis* is its spores. In nature, disease caused by this germ is usually associated with grazing animals such as sheep, goats, and cattle, and wildlife such as deer, wildebeest, buffalo, and elephants, which acquire the spores as they feed on vegetation. Some carnivores and omnivores, such as dogs, lions, and swine, may be susceptible through consumption of meat from infected animals. Many animals, such as birds, amphibians, reptiles, and fish, are not directly susceptible to anthrax. *B. anthracis* is found in soils throughout the world in Asia, Africa, South and Central America, the Caribbean, the Middle East, and parts of Europe, but it also can be found in certain U.S states, primarily along old cattle trails, such as in Texas, Louisiana, Mississippi, Arkansas, New Mexico, Oklahoma, and some Midwestern states. However, anthrax disease is rare in the United States, because such disease is controlled in animal populations by vaccination programs. Spores tend to be favored when the pH of a richly organic soil is higher than 6.0 and rainfall gives way to drought conditions. When herbivorous animals contract the infection, they can transmit the infection to humans by direct contact with animal products, such as hair, wool, hides, bones, etc. Of course, those in certain occupations are at increased risk for contracting anthrax, such as animal handlers, agricultural workers, veterinarians, and the like. About ninety percent of all human anthrax cases reportedly occur in millworkers handling imported goat hair. In fact, anthrax is also known as the "wool sorters' disease." Human disease may be introduced in three different ways: via the skin (cutaneous) through scratches or abrasions, by inhalation of spores, or by eating contaminated insufficiently cooked meat or meat products. Records of the disease date back to the Egyptians of 1500 B.C. Anthrax caused the "fifth" and "sixth" plagues of ancient Egypt. The word *anthracis* actually comes from the Greek word meaning "coal" and was so named because the germ can cause a black scab (eschar) to form on the skin of cutaneous anthrax victims. It is also sometimes referred to as the "black carbuncle" for a similar reason.

Anthrax is considered to be the single greatest biowarfare threat because it is easy to cultivate and spores are readily produced, although production of a weaponized spore is not so simple. Reference is often made by the media to "weapons grade" anthrax versus "non-weapons grade." Simply speaking, the term *weapons grade* relates to four things:

1. small spore size, usually one to three microns (a micron is a millionth of a meter)
2. lack of clumping (usually accomplished by addition of some polymer or other agent that prevents the natural tendency of spores to clump together)
3. amount of spores present
4. an effective delivery system

In this regard the anthrax unleashed on the U.S. in the fall of 2001 was near weapons grade—that is, it fulfilled two of the four criteria. It was of small particle size, it dispersed well in the air, but it was present only in limited quantities and it was delivered via the mail. If enough product had been made and if it had been effectively delivered, it would indeed have been considered weapons grade.

In nature, spores in soil tend to clump together, making it very difficult for a person to contract inhalation anthrax naturally. In order for a terrorist to weaponize anthrax spores, first he must prevent clumping of particles and then he must deliver the weapon in sufficient quantity. Clearly both of these tasks are very difficult. Even a crop duster would have to be retrofitted quite extensively in order to deliver spores effectively. Although these may be limiting factors for many individual terrorist cells or groups, they can easily be overcome by governments or heavily funded terrorist organizations. To be an effective weapon the anthrax spores must remain airborne in a concentration that is high enough to be inhaled deep into the lungs of victims. Based on the available data it appears that number varies from about eight thousand to forty thousand spores. A few studies suggest that inhalation of small numbers of spores, around 500 to 1,000 over an eight-hour period, did not make goat-hair millworkers develop disease. Although there are three types of infections that *B. anthracis* can cause (skin, inhalation, gastrointestinal), a bioterrorist would be most interested in causing inhalation anthrax because of the high mortality rate associated with it. In nature, however, approximately ninety-five percent of human cases of anthrax are cutaneous infections.

Anthrax spores can survive inside macrophage (one type of the body's phagocytic cells), eventually vegetating and growing to such numbers that they cause the cells to burst, thereby releasing bacilli into the bloodstream. These bacteria produce four virulence factors (factors that make germs cause disease), three of which are toxins (poisons) that cause the systemic symptoms observed, namely, protective antigen toxin, the lethal factor, and the edema factor. Lastly, they produce a capsule that protects them from destruction by the body's defensive phage cells. At this point, when toxins are released into the bloodstream and systemic symptoms occur, antibiotics are useless, because they have no effect on the toxins already made.

CLINICAL FEATURES

Inhalation anthrax

This form is really a biphasic disease in that the *initial* phase is characterized by mild flu-like symptoms followed by a period of apparent wellness for a day or so, immediately followed by the *acute* phase, which eventually leads to more serious symptoms. The incubation period can vary from one to five days, depending upon the number of spores inhaled, but can be as long as sixty days. The symptoms during the *initial phase* are characterized by the following: a nonspecific respiratory illness relatively mild in nature, myalgia (muscle pain), malaise, fatigue with a low-grade fever, and a nonproductive cough. Patients sometimes complain of a mild chest discomfort. The *acute phase* develops usually after two to five days and characteristically includes the following: an acute respiratory distress, difficulty breathing, profuse sweating, cyanosis (turning bluish in color), high temperature, and increased pulse and respiratory rate with chest sounds. If an X-ray is performed on the patient at this point, a mediastinal widening (swelling of lymph nodes under the breastbone) is very characteristic. Shock and death usually follow twenty-four to thirty-six hours after the respiratory distress onset. Fatality rate of inhalation anthrax approaches ninety percent even with antibiotic therapy. However, this figure will probably change for the better, owing to the availability of newer antibiotics and superior intensive-care treatment facilities. Inhalation anthrax is not spread via person-to-person transmission.

Cutaneous anthrax

This skin form of anthrax results after spores are introduced beneath the skin by inoculation or a contamination of a preexisting lesion or break in the skin. The incubation period ranges from two to seven days (rarely after one day) but more often occurs between two and five days. Initially the skin lesion starts out as small, painless, itchy pimples on some exposed part of the body such as the face, neck, or arms. This lesion becomes vesicular (like a blister), and afterward a small ring of vesicles may develop. These can join together (coalesce) into a single large vesicle that eventually ruptures to form an open ulceration. The ulceration eventually develops a black scab (eschar) at the center (within two to six days). The area of the lesion may become markedly endematous (swollen). The eschar falls off after about one to two weeks and leaves a permanent scar. Many systemic symptoms may also occur, such as fever, myalgia, and regional swelling of lymph nodes in the area that drains the site. In some cases significant swelling may occur (called malignant edema), which can interfere with the trachea (windpipe) if head and neck lymph nodes are involved. Septicemia (blood infection) is rare but can occur. When malignant edema occurs, a generalized toxemia can also result. Untreated cutaneous anthrax can have a fatality rate of up to twenty percent but fatalities are rare (less than one percent) with proper antibiotic treatment. While anthrax is not transmissible person-to-person direct exposure to vesicle secretions of cutaneous anthrax lesions can result in a secondary cutaneous lesion. Secretions from vesicles can be quite prolific so some caution is advised.

Gastrointestinal anthrax

This form of anthrax is caused by the ingestion of contaminated meat, in particular raw or undercooked, from infected animals. Up to the time of this writing there has never been a case of gastrointestinal anthrax reported in the U.S. The incubation period is much like that of cutaneous anthrax, ranging from two to seven days. There are two general types of GI anthrax, *intestinal* and *oropharyngeal,* each with a different set of symptoms. Intestinal anthrax generally presents initially as a nonspecific syndrome with symptoms of nausea, vomiting, loss of appetite, and fever. The disease progresses with an increasing abdominal pain and vomiting of blood (hematemesis), and a bloody diarrhea can develop, which can be accompanied by fluid buildup outside the intestines in the abdominal cavity (ascites). Oropharyngeal anthrax presents with swelling of the neck (edema), and a lesion can be seen in the oral cavity much like that of a cutaneous lesion, sometimes on the tonsils. Other symptoms include fever, swollen lymph nodes, and inability to swallow. Shock and toxemia can characterize both forms of the disease, especially in the terminal stages. The fatality rate for GI anthrax ranges from twenty-five to sixty percent.

Any one of the anthrax diseases discussed—inhalation, cutaneous, and GI—can be complicated by meningitis, which occurs in about five percent of the cases when anthrax bacilli enter the central nervous system via the bloodstream eventually getting through the blood-brain barrier.

Diagnosis

In part diagnosis will be a clinical assessment made by a physician or clinician after examining the patient, in conjunction with laboratory findings. Obviously many germs cause flu-like symptoms or cause wound lesions; that is why a laboratory assessment is so important.

- *Culture* is the definitive test for anthrax. *B. anthracis* can be isolated from specimens such as blood or spinal fluid or secretions from a wound or ulcer. Several of these may be taken by your physician.

- *Nasal culture:* We have been inundated with media reports concerning the significance of nose-culture tests. There is a great deal of confusion related to the usefulness of this type of test. Simply put, nose cultures are done for epidemiological purposes, not for diagnosis. In other words, they are useful to determine the degree of exposure to spores by a population in a particular area, such as the nose cultures done at the Trenton post office and Senator Daschle's office. By using nose cultures one can understand the extent to which anthrax spores spread in that particular population and area. This information may be used to determine whether prophylactic antibiotics should be provided to those exposed. However, the specific nose culture on any one individual does not determine a diagnosis for that individual. A negative culture may still result in treatment if culture positives were found in other individuals within the specific population tested. A negative nasal test therefore does not indicate that an individual is free of anthrax bacilli. For example, a patient may have a negative nose culture and yet may have been exposed to spores which have lodged deep within the lungs and which wouldn't be detected by nasal testing. There is no screening test currently available for the detection of anthrax infection in an asymptomatic person.

- *Microscopic* examination is an alternative test procedure to culture. A simple gram stain will tell whether a gram-positive (blue-staining) bacillus is present and thereby raise suspicion for the anthrax germ. Sometimes the anthrax bacterium can be seen under the microscopic in a vesicular lesion, but most times more sophisticated tests will have to be performed.

- *Special tests:* Sometimes the only way to identify the anthrax bacterium is by a special test called biopsy, where an immunohistochemical slide test shows evidence of the organism's destruction of that infected tissue. Another type of test, called the PCR (polymerase chain reaction), amplifies the bacterial DNA of the anthrax germ in the site so that a diagnosis can be confirmed. Lastly, there are numerous blood tests that can determine whether an individual has developed antibodies to the anthrax bacillus, thereby confirming exposure. These tests can be either enzyme-based, called ELISA (enzyme-linked-immunosorbent-assay), or they can be fluorescent tests called FA (fluorescent antibody) tests. Keep in mind that blood tests such as these can confirm an anthrax diagnosis but they may take a day or more to complete and may require several days or longer to develop sufficient antibody for testing. Scientists can also use both ELISA and FA tests to directly detect *B. anthracis* itself in a body site, but such testing can be performed only in special reference laboratories at this time.

Treatment

While all the treatment regimes presented require a prescription from a physician, they are being provided for informational purposes and to allay fears brought on by the erroneous assumption that only one antibiotic—cipro—is useful. Nothing could be further from the truth. There are many alternatives. Below are the recommended dosages for adults. Obviously the dosages for children would be different and would depend upon their ages and weights.

- *Quinolone Antibiotics*
  Ciprofloxacin   500mg twice daily
  Levofloxacin   500mg once daily
  Ofloxacin   400mg twice daily

- *Tetracycline Antibiotics*

  Doxycycline   100mg twice daily

- *Penicillin Antibiotics (for penicillin-susceptible strains°)*

  Amoxicillin   500mg every 8 hours

  Penicillin V   30mg/kg/day in 4 divided doses

  Penicillin G   2–4 million units every 4–6 hours

  °Penicillin antibiotics must be used with caution because some strains of *B. anthracis* possess an inducible penicillinase (an enzyme that inactivates penicillin) and may therefore become resistant to penicillins during treatment. However, penicillins are still recommended for prophylaxis.

- *Other Antibiotics*

  There are numerous other antimicrobial agents that can be used as alternatives in the event of an emergency if the above listed drugs are unavailable or in short supply. Among them are erythromycin, imipenem, clindamycin, vancomycin, and chloramphenicol.

The duration of post-exposure prophylaxis or treatment of inhalation anthrax is sixty days. It is currently recommended that any patients being treated for inhalation anthrax also should receive anthrax vaccine.

After three doses of vaccine are administered, at days zero, fourteen, and twenty-eight, antibiotics may then be discontinued. Currently full vaccination against anthrax consists of six doses at zero, two, and four weeks, then six, twelve, and eighteen months, followed by yearly boosters. According to health authorities there is limited human data that suggests that the vaccine would be effective against cutaneous anthrax; however, there is insufficient data on inhalation anthrax. Nevertheless, in rhesus monkeys it appears that there is good protection for up to two years. Currently the vaccine is not recommended for widespread use on the general public.

PROTECTIVE RESPONSE STRATEGY

• Probably the single most important protective measure that anyone can employ for themselves, in general, and, specifically, in light of the recent cutaneous anthrax cases, is hand washing. It is imperative to wash hands after handling mail. Hand washing should become an automatic health habit, because it can effectively dilute any contamination, including anthrax spores. When one considers that eighty percent of all infections are transmitted by contact, either directly or indirectly, hand washing should be performed several times during the course of the day. At a minimum you should do so before eating or drinking, after using a bathroom facility, and after contaminating your hands with a cough or sneeze or any work-related activity. It's also advisable to wash your hands after shaking hands with someone else. I also personally wash my hands whenever I come into the house from outside. Hand washing is an effective, simple, and underutilized health-maintenance technique.

• Various governmental bodies distributed some do's and don'ts regarding handling the mail. The following is a summation list of those recommendations:

–Common sense must prevail at all times.

–First examine unopened envelopes and packages for foreign bodies or powder.

–Don't open letters with your hands; use a letter opener.

–Open all envelopes and packages carefully and slowly with a minimum of movement.

–A suspicious letter may have threatening messages written on it; may have a strange odor; may be soiled or oily; may have an envelope with no return address; may be an unexpected envelope from a foreign country; may have an envelope that contains improper spelling of common names, places, or titles.

- Open mail in a designated place. I open mine on a polymer cutting board with built-in germicide protection. These days every caution is indicated. Open mail slowly and carefully so that its contents are not disturbed. After the task is completed, sanitize the board and area around the board with a 1:10 dilution of household bleach (take one part 5.25% hypochlorite bleach and add nine parts water). Let the bleach remain in contact for at least thirty minutes before rinsing. Be sure to use gloves. Ordinarily a 1:10 bleach solution (ten percent bleach solution) is an excellent germicide. It effectively kills most germs very well, even hearty *Staphylococcus aureus* and *E. coli* bacteria, as well as viruses like those that cause AIDS or the common cold and flu. However, it does not ordinarily kill spores completely. In the event that it becomes necessary for you to decontaminate a surface that might contain numerous spores of anthrax, concentrated bleach should be applied (5.25% hypochlorite) for at least thirty minutes. Extreme caution is indicated here because concentrated bleach is a dangerous and powerful corrosive chemical. This later method should be applied only in an emergency situation.

- Never bring envelopes close to one's face.

- Immunocompromised people should have their mail opened by a relative or friend.

- In high-risk areas, respiratory mask N95 could be worn for protection when opening mail.

- If an article of mail is suspect, put it in a baggie with a zip lock or put a piece of plastic wrap over it, or a wet towel or cloth, then weigh it down with a book or magazine. Call proper authorities. Spores tend to adhere to the plastic because of static electricity, and a wet towel contains spores. If the post office decides to use an irradiation, ozonation, or some other technique to permanently sanitize the mail, this strategy may become unnecessary.

- Clothing may be a source of contamination. Wash-and-wear clothes should be washed with detergents that contain sanitizing agents in them. Whenever possible, hot water (at least 155°F) should be used. Periodically run an empty cycle with bleach and water in order to disinfect the washing machine itself. Finally, if climate and other circumstances allow, consider line-drying wet clothes in the sun rather than putting them into a dryer. The sun's UV rays do a wonderful job of killing germs that may be lingering on the laundry.

- Any cut, scratch, or abrasion should be covered with a dressing that has been coated with one percent tincture of iodine. If the cut or abrasion is not too extensive, apply tincture of iodine to its surface. Iodine is an outstanding antiseptic and can be sporocidal. Always engage in proper wound care. No contusion is too small to warrant attention. Especially these days.

- If you have a wound that does not seem to heal or continues to drain fluid, or if your wound is abnormally swollen (with or without a scab), consult with your doctor.

- Homes that have central air-conditioning systems fitted with HEPA (high-efficiency particulate air) filters will remove more than 99.97 percent of particles larger than 0.3 microns in size. Since anthrax spores are about 1 micron in diameter and tend to clump in twos or threes, making the size range from 1 to 3 microns, they will be efficiently trapped.

- Perhaps one of the most important **Protective Response Strategies** is to be alert and aware of circumstances around you. If you see any anomaly or anything that appears strange, report it to the police or to 911. You may save someone's life or even avert another tragedy.

2.
PLAGUE

Bioterror Agent	*Yersinia pestis*
Type of Weapon	Bacterium
Disease Caused	Plague
Transmissible (person-to-person)	Yes
Incubation Period	2 to 10 days
Cardinal Features of Disease	*Bubonic:* swollen buboes (groin lymph), high fever *Pneumonic:* high fever, cough, chest pain, hemoptysis (coughing up blood)
Treatment	Antibiotics: tetracycline, quinolones, gentamicin
Vaccines Available	Yes

HISTORY AND BACKGROUND

Most of us are familiar, at least in passing, with the Black Death or the plague that befell Europe in the fourteenth century. This disease killed more than twenty-five million people, which was calculated to be twenty-five percent of the entire population of Europe. This period is called the "second" pandemic. The first of three pandemics actually started in A.D. 541 and continued through the eighth century with an estimated forty million deaths. The third and last pandemic began in China in the 1860s and spread to Africa, Europe, and the Americas. The plague is actually a zoonotic disease that primarily infects rodents. The disease is caused by a bacterium called *Yersinia pestis* and is transmitted by a rat flea (*Xenopsylla cheopis,* the Oriental flea, or *Pulex irritans*, the human flea). In the past, epidemics in humans have originated by contact with fleas of infected rodents. The disease in humans has three clinical forms:

1. bubonic, characterized by swelling of the lymph nodes (buboes)

2. pneumonic, in which the lungs are extensively involved

3. septicemic, in which the bloodstream is infected with the *Yersinia pestis* bacterium.

* * *

Two epidemic forms exist: urban plague, which is maintained by the urban rat population, and sylvatic plague, which is endemic in the western United States and is carried by wild mice, rabbits, skunks, moles, gerbils, prairie dogs, and rats. In the past the great plagues were a result of the spread from rats to man in crowded unsanitary urban areas. Man is only accidental to the usual cycle: rodent-flea-rodent. In the United States most naturally occurring cases of plague in humans are concentrated in two regions: (1) the Southwest (Arizona, New Mexico, Utah, and Colorado); and (2) the Pacific States (California, Oregon, and Nevada). About a dozen cases of human plague are reported annually in the U.S. The chief enzootic areas are India, Southeast Asia, Africa, and North and South America.

Yersinia pestis is a short, rod-shaped gram-negative (stains red with aniline dyes), bipolar-staining (has a darkened area on either end resembling a safety-pin form) bacterium.

In a biowarfare scenario, the plague bacillus could be delivered by contaminated fleas as vectors causing the bubonic plague or, more likely, by an aerosol spread causing pneumonic plague. In nature a rat flea regurgitates the bacterium upon biting its host. Cats are also susceptible and, in fact, can themselves transmit the pneumonic plague to human beings. Unlike anthrax, which cannot be transmitted person-to-person, pneumonic plague can be transmitted via large aerosol droplets, usually from a coughing patient. In humans the mortality of untreated pneumonic plague is one hundred percent; the mortality of untreated bubonic plague is about fifty percent.

Yersinia pestis bacteria can be killed by the body's phagocytic cells called polymorphonuclear cells, but they can survive in the body's other phagocytic cells called monocytes, where they produce a capsule (a thick outer covering) called F-1 (fraction 1), which enables the bacteria to resist phagocytosis, or ingestion by white blood cells. These bacteria can then rapidly reach the lymph system and can enlarge the lymph nodes, causing hemorrhage. The lymph nodes may undergo a process called necrosis (death of areas of tissues), whereupon the bacteria can reach the bloodstream and become widely disseminated. Hemorrhagic and necrotic lesions can develop in all organs in the body.

It is thought that less than one hundred organisms are necessary to cause human infection. Studies have shown that the bacteria can remain alive in the soil for up to one year and up to 270 days in live tissue. The bacterium is killed by fifteen minutes' exposure to heat of $72°C$ ($160°F$), and dies within several hours of exposure to sunlight.

CLINICAL FEATURES

Bubonic plague

The incubation period ranges from two to ten days. Malaise, high fever, and tender lymph nodes (buboes) are characteristic. As the name implies the bubo or inguinal (groin) lymph nodes enlarge, but cervical (neck region) and axillary (under the arms) can also be involved. The nodes can become very tender and fluctuant (compressible). They become necrotic over time. In bubonic plague, victims may suffer from the septicemic form, and the bacteria can also spread to the central nervous system, the lungs (this produces the pneumonic disease, which can then be spread person-to-person), as well as elsewhere. Vomiting and diarrhea may develop with early sepsis. Later, shock, renal failure, and heart failure may occur, leading to death.

Pneumonic plague

The incubation period is somewhat shorter for pneumonic compared with bubonic plague. It's generally one to three days. High fever, cough, and chest pain are prominent features. Production of a bloody sputum (hemoptysis), headache, chills, malaise, myalgia, and evidence of bronchopneumonia also are characteristic symptoms. A sputum rich with gram-negative rod-shaped bacteria is either watery or muco-purulent (rich with mucus and pus). As the pneumonia progresses, difficulty in breathing as well as cyanosis (turning blue in color) occur. The terminal event involves both a respiratory and circulatory collapse.

Diagnosis

Certainly clinical assessment made by a physician or clinician after examining the patient is important for diagnosis, but this must be done in conjunction with laboratory findings. Although some of the symptoms are characteristic for plague, the initial symptoms mimic other diseases, reemphasizing the importance of laboratory testings.

- *Culture* for *Yersinia pestis* can readily be done from sputum, blood, and lymph aspirates. The plague bacteria grow on simple sheep-blood agar plates but do so slowly. Culture is useful for confirmation of the gram stain as well as other rapid and more special tests.

- *Microscopic:* A gram-stained smear of sputum will reveal a characteristic gram-negative safety-pin bipolar-staining bacillus.

- *Special tests:* Fluorescent antibody tests can be performed on respiratory secretions to detect *Yersinia pestis*. PCR tests can also be performed, but both procedures can be done only at specialized regional laboratories. Capsular F-1 antigen detection can also be performed by an immunoassay procedure performed on blood (serum).

Treatment

As is true for anthrax, *Yersinia pestis* can be successfully treated with many different antibiotics. Tetracycline, streptomycin, gentamicin, chloramphenicol, and quinolone antibiotics are highly effective especially if initiated within twenty-four hours of the onset of symptoms. Treatment is recommended for ten to fourteen days. Although the typical patient is afebrile (without fever) in only three days, an extra week or more of therapy prevents relapses.

Recommended post-exposure prophylaxis for exposure to *Y. pestis* is as follows:

- *Tetracyclines*
    Doxycycline   100mg twice daily
    Tetracycline   500mg four times a day

- *Quinolone Antibiotics*
    Ciprofloxacin   500mg twice daily

Prophylaxis medication need only be given for seven days.

There is a formalin-killed vaccine that is available in the United States, and in fact it has been extensively used. It seems to be effective against the flea-borne plague (bubonic), but it has not yet proven to be effective for the pneumonic plague. A measurable immune response is attained after three doses at zero, one, and four to seven months. Boosters are required every one to two years. Currently live-attenuated

vaccines are being tested. Post-exposure immunization has no utility.

PROTECTIVE RESPONSE STRATEGY

- A protective strategy for the bubonic and
  pneumonic plague is really a three-pronged
  procedure. Any individual needing such protection should don a surgical mask, spray an
  insect repellent on his or her clothes as well as
  exposed skin, and practice good hand-washing
  techniques. Each of these strategies is
  described in detail below.

- Since patients with pneumonic plague are
  infectious and transmit infection by large particle droplets (greater than five microns) gen-
  erated by coughing, talking, or sneezing, a
  simple surgical-type mask can protect workers
  or family members. Hence these "droplet precautions" should be maintained until the
  patient has completed at least three days of
  antibiotic therapy, after which a person is no
  longer contagious.

- Since "droplets" usually occur only within
  three to five feet of a patient, central air-
  conditioning systems do not need to be fitted
  with HEPA filters in order to prevent spread
  of infection.

- If a bioterrorist outbreak of plague occurs, control of fleas, rats, and other animals like stray cats is required. Insecticides and repellents are widely available.

- Unquestionably hand washing is just as important for plague bacilli as it is for anthrax prevention and should be performed as already outlined.

- Interestingly, since *Y. pestis* does not have a spore form, surfaces can be decontaminated with a simple germicide like ten percent bleach solution (one part bleach in nine parts water). Skin can be decontaminated using any good germicidal soap product.

- Although re-aerosolization of *Y. pestis* from contaminated clothing of persons exposed is low, clothing should be washed using a detergent containing a germicide. Hot water (155°F) should be used during the washing process. Clothes should also be dried using a hot cycle.

- Stand clear (farther than three to five feet) of individuals taking a coughing fit or spitting up sputum with or without blood.

3.

SMALLPOX

Bioterror Agent	Variola major
Type of Weapon	Virus
Disease Caused	Smallpox
Transmissible (person-to-person)	Yes
Incubation Period	10 to 14 days
Cardinal Features of Disease	Flu-like symptoms, then skin rash to pimple to vesicle (blister) to pustular (pus-filled) to crustular (scabs) lesions Lesions prominent on face and extremities
Treatment	Vaccine and immune globulin; cidofovir
Vaccines Available	Yes

HISTORY AND BACKGROUND

Smallpox has the dubious honor of being the greatest single killer in the recorded history of the human species. About 500 million people have been killed by this virus through the years. Chinese folklore alleges that smallpox vaccination occurred as early as the sixth century, although the first written record appears to be attributed to a Buddhist nun practicing during the reign of Jen Tsung (A.D. 1022–1063). She recommended selecting scabs from victims and letting them dry for about a month, then grinding them with particular plants and blowing the mixture into the nostrils of healthy individuals. Similar methods were employed by sixteenth-century Hindus in India. But notwithstanding these historical anomalies, the British physician Edward Jenner gets the credit for the first scientific attempt to control an infectious disease by means of systematic inoculation (vaccination).

In 1796, Jenner demonstrated that individuals injected with cowpox (a disease, similar to smallpox, that infects cows but not people) were protected against smallpox. Years later, Jenner admitted that he had been inspired to do this by reading about an English cattle breeder by the name of Benjamin Jisty, who became immune to smallpox after contracting cowpox from his herd, and who then inoculated his wife and two children with cowpox in order to avoid a smallpox epidemic. The technique that Jenner employed in his experiments came to be known as vaccination, from the Latin word *vacca*, meaning cow. Louis Pasteur actually coined the word in honor of Jenner, after applying his concept of injecting a weakened form of a pathogen into an animal, which thereby gained immunity against the more virulent form of the germ.

The smallpox virus, aka variola, is a member of the Orthopoxvirus group of viruses. There are two variants of smallpox: variola major, which is associated with a higher mortality rate of fifteen to forty percent, and variola minor, which causes a milder disease and is associated with a mortality rate of only one percent. Smallpox is the human type of poxvirus. There are other poxviruses that naturally infect animals, but they also can cause incidental infection in humans (zoonoses). These viruses share common antigens with smallpox, thus allowing them to be used as vaccines for humans. In fact, the vaccinia virus has been the historically chosen animal virus for vaccine production. Vaccinia viruses are slightly different from the cowpox virus, which is thought to be its ancestor. The variola viruses have a very narrow host range—primarily humans and monkeys—whereas vaccinia viruses have a broad range that includes rabbits, mice, and other animals. Using the vaccinia virus as a smallpox vaccine consists of simply scratching that live virus into a patient's skin. If the vaccine "took," then vesicles would develop at the scratch site. These would give way to pustules. It is important to stress an important point about smallpox disease vaccinations. Many complications related to vaccination occurred as smallpox waned. There were more complications related to the vaccinations than there were cases of smallpox. Some of these complications were severe, including encephalitis, and in immunocompromised patients who were vaccinated, sometimes inadvertently, fatal reactions were reported. We learned one other important point. Vaccinia viruses are relatively easy to spread from person-to-person among unvaccinated immunocompromised patients in close contact. This would still be the case if a massive national vaccination program were to be considered. There would be a large group of immunocompromised patients that simply would not be candidates for the vaccine.

In 1967, the World Health Organization introduced a worldwide campaign to eradicate smallpox. At the time there were thirty-three countries with endemic smallpox and about fifteen million cases per year. The last Asiatic case occurred in Bangladesh in 1975; the last natural case was diagnosed in Somalia in 1977. The last human case occurred in 1978, which was a laboratory infection. Smallpox was officially declared eliminated in 1979, and variola virus stocks have been destroyed in all laboratories in the world except for two, which are World Health Organization collaborating centers—one is in Atlanta, U.S.A., the other is in Moscow, Russia. Currently there is a debate as to whether these remaining stocks of small-pox should be destroyed. If they were and if any cases of smallpox were to appear after such destruction there would be no ability to do further research or study of this germ. In spite of its number-one standing as the all-time killer of man we still don't quite understand what made the variola virus so virulent! Also the potential weaponization of smallpox continues to pose a significant threat to the free world by potential bioterrorists or rogue governments. We may need the smallpox to counter such a threat.

Smallpox is very contagious and as such it is similar to the plague germs in that it is highly transmissible from person to person. It is currently assumed that an aerosol infective dose is low and presumably ranges from ten to one hundred organisms.

CLINICAL FEATURES

The portal of entry of smallpox virus is the mucous membranes of the upper respiratory tract. Smallpox is transmitted by either large or small respiratory droplets, and by contact with skin lesions or secretions. Patients are considered more infectious if they are actively coughing. The following scenario is thought to take place:

• the virus multiplies in the lymph tissue that drains the site of entry

- the virus then makes its way to the blood-stream, causing infection of the blood (viremia)

- there is a second multiplication that takes place in the spleen, liver, and other lymphatic tissues

- this leads to a secondary viremia

- finally clinical disease occurs

The incubation period is typically twelve days with a range of ten to fourteen days (the exception would be the Russian strains, which would have an incubation period range from three to five days). The clinical illness begins with a two-to-three-day period of vague symptoms such as malaise, fever, headache, chills, and backache. The fever can last as long as five days or as short as one day. Usually after the fever an exanthem (eruption of skin or rash) appears, which undergoes the following sequence of events. It begins as a papular (pimple) lesion for one to four days, then it becomes vesicular (blisterlike) for one to four days, and then pustular (filled with pus) for two to six days, which forms crusts (crustular) that fall off in two to four weeks after the first skin lesion appears, leaving pink scars. Because there are so many exanthems that occur with other diseases it is imperative to observe the lesions very carefully. An important characteristic of smallpox lesions is that all lesions in any affected area are generally found in the same state, which is unlike chicken-pox lesions. Chicken-pox lesions on the other hand are not synchronous; instead they occur in crops. Smallpox lesions are also said to be distributed centrifugally (more numerous on the face and extremities rather than the trunk), unlike chicken pox. Hence the exanthem for smallpox is very characteristic and is an important differential criteria for diagnostic purposes. The case fatality rate in unvaccinated patients is fifteen to forty percent. In vaccinated people the fatality rate is less than one percent.

Patients with smallpox are infectious as soon as a rash is evident and remain infectious until their scabs fall off, which is approximately three weeks in duration. There is a rare form of smallpox called "hemorrhagic variola" that is very pathogenic and has a very high mortality associated with it.

Diagnosis

Because there are numerous exanthemous diseases, both infectious and noninfectious in nature, smallpox must be properly differentiated from them. However, the clinical exanthemous lesion sequence is quite unique, so a clinical assessment can easily be made by a competent physician.

- *Culture:* Skin lesions are the specimen of choice for viral isolation. Since viruses cannot grow on artificial agar media, such as that used to grow bacteria, living cells must be cultured. Viruses need living cells in order to grow. Alternatively smallpox can be grown in chick embryos. Nasal swabs and respiratory secretions can also be used as specimen types to culture for smallpox. Serum (blood) can be successfully cultured also.

- *Microscopic* examination using an electron microscope can be used to identify the characteristic morphology of the smallpox virus very rapidly, usually within an hour or so. The problem with this test is that not many laboratories are equipped with electron microscopes.

- *Special tests:* a PCR test can be used to examine respiratory secretions in much the same way as for anthrax and plague bacilli. There is an agar gel precipitation test that can be performed on a skin lesion that can identify smallpox relatively quickly. Because antibodies to the smallpox virus appear after the first week, a serology test can tell whether an individual has developed antibodies. However, as with most serology tests this is really a retrospective analysis used to help confirm diagnosis.

Treatment

Vaccinia immune globulin (VIG) must be used in conjunction with a vaccinia vaccine if an individual's exposure to a smallpox case was more than four days earlier. However, less than four days after such contact, only a vaccinia vaccine is required. The vaccine starts to be protective in about seven days but doesn't confer life-long immunity. Revaccination is recommended at five-to-ten-year intervals. Vaccination against smallpox is contraindicated in patients with skin disorders like eczema and in immunosuppressed or immunodeficient patients, because of the possible development of severe life-threatening complications. Vaccination is performed by a unique process called scarification, where an intradermal introduction of the virus is made by scratching the skin. Interestingly, there are some antiviral drugs such as Cidofovir that have demonstrated that they confer some protection against infection. Unfortunately, because smallpox has been eradicated there has been limited research on such drugs.

PROTECTIVE RESPONSE STRATEGY

- How would a terrorist likely introduce small-pox on a population? There are two primary methods of spread: aerosol dispersal and contact. Terrorists would presumably prefer an aerosol delivery system, as this would expose the largest number of people to the germ. However, this is not so easy to do. The second delivery method, direct contact, could involve contaminating a group of volunteer terrorists who then infect large numbers of individuals as they interacted over a period of days or weeks. It is important to realize that if a smallpox outbreak is detected in an area, the CDC and other health authorities would be able to contain the outbreak by a program that would involve immediate vaccination of all individuals surrounding the index (first) case or cases in a series of concentric rings progressing outward until all desired vaccinations have been accomplished. Remember that post-exposure immunization with smallpox vaccine (vaccinia virus) is effective and is recommended if given within four days of exposure. However, even if more than four days has elapsed since exposure, both vaccination and vaccinia-immune-globulin can be given in order to provide protection. That was precisely the method used to finally eradicate smallpox from the face of the earth, which should be just cause for some optimism.

• If a major outbreak of smallpox occurs, be prepared to stay indoors for a period of a few days to weeks in order to reduce contact with possible contamination and to contain the epidemic. You should maintain a few weeks' supply of food in your household.

- Generally, respiratory droplets of smallpox cases are infectious earlier than skin lesions, but caution is in order for both sources of the virus (droplets and contact). In hospitals, both airborne and contact precautions are instituted to prevent contagion. However, unlike for plague precautions, a simple surgical mask is insufficient protection. A special respirator mask called the N95 is recommended and must be worn for such protection. Smallpox patients should be quarantined from the time the rash first appears until the scab finally falls off (about three weeks).

- If a household member has contracted smallpox, all clothing, bed linens, and other materials that touched the patient must be decontaminated with a germicide such as a 1:10 bleach solution, or steam, or heat (285°F for 180 minutes or 340°F for one hour). Under refrigeration the virus maintains its infectivity for many months.

- Hand washing with a good germicidal soap is a must whenever any kind of contact is made with a smallpox patient or the patient's environment. See comments previously made.

- Rest assured that the post-9/11 anthrax attack has caused health facilities and city, state, and federal authorities to prepare for just about any catastrophic event, even a smallpox attack.

4.

BOTULINUM TOXINS

Bioterror Agent	Botulinum toxins of *Clostridium botulinum*
Type of Weapon	Bacterial toxins
Disease Caused	Botulism
Transmissible (person-to-person)	No
Incubation Period	18 to 72 hours
Cardinal Features of Disease	Double vision, lack coordination of eye muscles, inability to swallow or speak, generalized weakness and dizziness, respiratory paralysis
Treatment	Antitoxins
Vaccines Available	Yes

HISTORY AND BACKGROUND

Clostridia are anaerobic (live in the absence of oxygen), gram-positive, spore-forming bacilli that morphologically (shapewise) resemble *Bacillus anthracis,* except that *B. anthracis* is an aerobic bacterium. Among its pathogenic types are organisms that can cause botulism, gas gangrene, and tetanus. The clostridia bacteria cause the aforementioned diseases by the protein toxins that they produce. *Clostridium botulinum,* which causes botulism, is ordinarily found in soils worldwide, and sometimes it can also be found in animal feces. During the growth of *C. botulinum,* toxins can be liberated into the environment. There are seven toxins that are produced (A, B, C, D, E, F, and G) by different strains. The principal causes of human illness are produced by types A, B, E, and F. Types A and B are mostly associated with a variety of foods, while type E is mostly associated with fish products. Type F is the least common. Types C and D are associated with botulism in birds and mammals, but not humans. Type G organisms have been isolated from autopsy samples from a number of individuals who have died suddenly, but it is unclear as to whether they actually cause botulism. The toxins act as neurotoxins, each in the same way no matter the type. They act primarily by binding to synaptic vesicles of nerves, thereby preventing the release of acetylcholine at the peripheral nerve endings, and patients develop what is called acute flaccid descending paralysis. These toxins thus block neurotransmission. This interruption of neurotransmission causes both palsies as well as skeletal muscle weakness, which are the primary clinical features. These toxins are proteins with molecular weights of about 150,000. *Clostridium botulinum* toxins are among the most toxic substances known to man. A lethal dose for humans is only about one microgram, which is one-millionth of a gram.

The CDC recognized that there are four different categories of botulism disease that occur in humans. The classic type is *food-borne botulism,* which typically occurs in adults and is caused by ingestion of toxin present in contaminated foodstuff. *C. botulinum* grows in the food and produces its toxin. The usual foods involved are canned alkaline foods that are eaten without cooking, or smoked and vacuum-packed foods. The usual scenario involves spores of *C. botulinum* growing under anaerobic conditions into their vegetative forms and then producing toxin. The second category is called *wound botulism,* which is considered to be the rarest kind of botulism. In this type of illness the bacterium gains access to a wound site and then produces its toxins in vivo after multiplying sufficiently. The third category is called *infant botulism,* and this form of botulism is the most common type of all. A child may consume a foodstuff that was contaminated with *C. botulinum* instead of consuming preformed toxins. The production of the botulinal toxin actually occurs in the infant's gut, and thus the infant is poisoned from within. The last category is called *classification undetermined,* which is really a catchall group. In this category are individuals older than twelve months in whom no food or wound source of bacterium can be found.

During a bioterrorist attack it is anticipated that botulinum toxins would be delivered as an aerosol, and this delivery would result in symptoms similar to those encountered with food-borne botulism. We know that botulinum toxins can be weaponized, because the Iraqi government admitted to a United Nations inspection team in August 1991 that it had done research on these toxins prior to the Persian Gulf War. Further, we know that they actually deployed over one hundred such weapons with botulinum toxin. It is possible for any of the seven known botulinum toxin types to be weaponized, as they all would theoretically have the same effect. A terrorist group would grow the *Clostridium botulinum* bacterium, allowing it to produce toxins. The group would then purify the toxins made and accumulate large quantities for a probable aerosol assault. Certainly it can also be weaponized by placement in foodstuffs, but delivery of the toxin in that manner would not be as efficient. Nevertheless, the intent of a terrorist might be to sabotage food supplies in a targeted event.

CLINICAL FEATURES

The symptoms of botulism begin from eighteen to twenty-four hours after ingestion or inhalation of the toxin, although several days may pass before symptoms occur. The initial symptoms include double vision, lack of coordination of eye muscles, inability to swallow, speech difficulty, generalized weakness, and dizziness. These symptoms are followed by a descending progressive weakness of the extremities along with weakness of the respiratory muscles. There is no fever and the patient may be totally alert and oriented. Neurological examination shows flaccid muscle weakness of the tongue, larynx, respiratory muscles, and extremities. A patient would remain fully conscious until shortly before death. Death occurs from respiratory paralysis or cardiac arrest. The mortality rate can be quite high. Interestingly, patients who recover do not develop productive antibody (antitoxin) in the blood. There is probably a suppression of antibody production caused by the toxins in much the same way that toxic shock syndrome toxin-1, produced by *Staphylococcus aureus,* prevents antibody production. In infants, the signs of paralysis are called "floppy baby" syndrome.

Diagnosis

In natural cases, routine laboratory findings are of no value in making the diagnosis. However, toxin may be demonstrated in a patient's serum. Certainly clinical symptoms are very characteristic and the differential diagnosis would initially involve ruling out various similar neuromuscular disorders such as Guillain-Barré syndrome, myasthenia gravis, tick paralysis, or even nerve gas exposure.

Nasal swabs and induced respiratory secretions for PCR analysis (which would assay for contaminating DNA from *C. botulinum* bacteria that would probably be present in toxin made by a terrorist) and toxin assays (which would detect preformed toxin). Also a blood sample can be taken for toxin assays. Both the toxin assay and PCR for DNA of *C. botulinum* would be performed by the CDC or a regional health department laboratory. There are no other test procedures that are useful to help diagnosis of botulinum toxin exposure. Exposure to botulinum is not contagious!

Treatment

Since the botulinum toxin blocks the action of nerves that activate muscles necessary for breathing, leading to death by suffocation, an antitoxin can be injected up to about twenty-four hours (based on monkey studies) after exposure to a lethal toxin dose and still prevent death. There are two types of antitoxins available—a trivalent (includes types A, B, E) and a heptavalent (types A, B, C, D, E, F, and G) preparation. These products have been prepared from equine (horse) species. As with all medication, there is a downside—

namely, there is a theoretical risk of serum sickness developing with administration of such antitoxins. However, that is a small price to pay when one thinks of the alternative! Other treatments would include mechanical respiration if necessary. Cases of botulism reported in the 1950s had a mortality rate of sixty percent; however, with tracheostomy and modern-day ventilation assistance, fatalities have been reduced to less than five percent or so. A vaccine preparation is also available that would allow an individual to develop antibodies to the five most common *C. botulinum* types: A, B, C, D, and E. The vaccine consists of the five common toxins rendered harmless as toxoids to which protective antibodies are produced. The current recommended schedule is to administer the vaccine at zero, two, and twelve weeks, followed by a one-year booster. Studies have shown that this vaccine regimen produces protective antitoxin levels in greater than ninety percent of those vaccinated. Again, as with all vaccines, there are some mild local reactions associated with the botulinum vaccine, but rarely are there any severe reactions. Reactions include localized redness and edema at the injection site as well as systemic symptoms like fever, malaise, headache, and myalgia.

PROTECTIVE RESPONSE STRATEGY

- Because individuals may not show any signs of illness from about eighteen to seventy-two

hours, victims have an important advantage—
time. If symptoms of double vision, inability to
swallow, speech difficulty, generalized weakness or dizziness occur, it is time to get to an
emergency room ASAP. Do not drive there
yourself; ask someone else to drive, or call 911
and advise the operator that you think that you
may have been exposed to a toxin, perhaps
botulinum. Other kinds of toxins are quicker-
acting. For example, a nerve gas would cause
reaction in minutes and a staphylococcal
enterotoxin B reaction would occur in a few
hours. (I'll discuss these agents in later chapters.)

- The fact that there are both antitoxins and vaccines available for treatment and prophylaxis is
encouraging.

- If a bioterrorist were to introduce botulinum
toxin aerosolly, the actual toxicity and lethality
would be *less* by inhalation than by a food-
borne assault.

- Always keep a supply of activated charcoal
tablets in the medicine chest, as they can
absorb almost all poisons by chemically binding the poisons to limit further harm.

- Soap and water can be very effective at remov-
ing most toxins from skin, clothing, and equip-

ment, so decontamination of a toxin is *not* as critical as decontamination of an infectious germ. Nevertheless, a mild bleach solution (one part bleach in nine parts water) would also effectively inactivate most protein toxins.

- A protective mask, if worn properly, is effective against toxin aerosols. However, it is important that a tight fit is achieved, since even a small leak could result in significant exposure.

- Keep in mind that because a toxin attack is not dermally (skin) active, special protective clothing, other than a mask, is not as important as it would be for a chemical attack. Ordinary clothing would provide a body shield from toxin and routine washing of clothing would effectively remove toxin.

5.

BRUCELLOSIS

Bioterror Agent	*Brucella melitensis, suis, abortus, canis*
Type of Weapon	Bacteria
Disease Caused	Brucellosis
Transmissible (person-to-person)	No
Incubation Period	1 to 6 weeks
Cardinal Features of Disease	Undulant fever (waxing and waning of fever), flu-like
Treatment	Antibiotics: doxycycline and rifampin
Vaccines Available	Yes

HISTORY AND BACKGROUND

The brucellae are small, nonmotile, aerobic, gram-negative short bacilli or coccobacilli. Brucellosis is a systemic zoonotic disease caused by any one of the four species of bacteria: *Brucella melitensis, B. suis, B. abortus,* and *B. canis.* These bacteria ordinarily cause disease in domestic animals, such as goats, sheep, and camels *(B. melitensis),* cattle *(B. abortus),* and pigs *(B. suis).* The primary pathogen of dogs, *B. canis,* rarely causes disease in humans. Natural infection in humans occurs when bacteria are inhaled as aerosols, or if raw (unpasteurized) infected milk or meat are ingested, or if they come in contact with abrasions in skin or with conjunctival surfaces. The disease in humans is historically called undulant fever, Malta fever, Bang's disease, Gibraltar fever, and Mediterranean fever. The organism was first isolated in 1887 by Sir David Bruce, who first discovered the bacterium from the spleens of British soldiers dying of Malta fever during the Crimean War. It was sometimes referred to as "Mediterranean gastric remittent fever" because of its symptoms, which I'll describe later. A Mediterranean Fever Commission was established in order to identify the cause of this disease. That commission identified goats as the cause of human infection on Malta and restricted the ingestion of unpasteurized goat's milk and cheese. Soon these foodstuffs were banned and soon the number of cases of brucellosis decreased. Subsequent to this discovery, several other animals were found to harbor a similar organism. The disease occurs worldwide, especially in the Mediterranean and Arabian countries, India,

Mexico, Asia, and Central and South America. In developed countries, some of human infection is associated with meat-packing and the dairy industry. Cases in the U.S. are also related to the animal-husbandry industry. The number of cases in the U.S. has continued to decline over recent years, primarily because of extensive vaccination programs of animals. In addition to zoonotic natural infections, numerous laboratory infections have occurred over the years. The aforementioned infections clearly demonstrated that human-to-human infection does not occur; therefore isolation of infected patients is not required. Several studies report that human brucellosis is underdiagnosed and underreported. It has been estimated that from twenty-five to thirty cases go unrecognized for every case reported. If *Brucella* were to be used as a biological warfare agent, it would likely be delivered via the aerosol route. The germs would move from their entrance point to the lymph channels and nodes, eventually reaching the thoracic duct and bloodstream. Abscesses can form in lymph tissue, liver, spleen, bone marrow, etc. The brucellae are intracellular bacteria. Once the organisms are phagocytized they are able to survive within these cells and as such they can be carried to the bloodstream and thereby deposited in multiple organs of the body. The bacteria grow inside the phagocytes and eventually kill their host cells, and a new crop of bacteria is released. The undulant fever pattern observed with this disease corresponds to the release of bacteria into the blood, thereby causing fever. As these bacteria are cleaned up, the fever subsides, only to recur when another crop of bacteria are released. Relapses are common. *Brucella* species have two morphologically different types of colonies: one type is "smooth," the other "rough." The smooth form is more pathogenic because of the presence of a capsule that protects the bacterium against being phagocytized and destroyed.

Of the four species, *B. melitensis* and *B. suis* are the more virulent species. In addition to their virulence these species also survive intracellularly better than other species. On the other hand, *B. abortus* and *B. canis* are insidious in their onset but tend to cause milder disease with fewer complications.

CLINICAL FEATURES

The incubation period ranges from one to six weeks but normally ranges from three to four weeks. The onset of brucellosis is insidious, with malaise, fever, chills, sweats, headache, fatigue, myalgias, and arthralgias. Fever usually rises in the afternoon and it falls during the night, and is accompanied by drenching sweat. Swollen lymph, spleen, and liver may also be present. The "undulant fever" can occur over weeks, months, or even years. Yet there are many days when a patient has no fever and feels relatively well, only to experience another cycle of waxing and waning fevers. Often physicians may categorize these patients as having fevers of unknown origin. Cough occurs in about twenty percent of cases but the X-ray appears normal. Lethality may approach about six percent if *B. melitensis* is the agent but less than one percent if any of the other species are involved. Most deaths are associated with endocarditis (infection of the lining of the heart) or meningitis (infection of the membranes around the brain). Gastrointestinal symptoms occur in up to seventy percent of adult cases (less frequently in children). Rashes occur in less than five percent of cases and include macules, papules, ulcers, and erythema.

Diagnosis

The initial symptoms of brucellosis are usually nonspecific and require a physician to rule out a large number of other infectious diseases that mimic it, including virus and bacteria. More important, brucellosis is indistinguishable from tularemia and typhoid fever. Both of these diseases will be discussed in upcoming sections.

- *Nasal swabs*, sputum, and induced respiratory secretions are useful for culture and PCR testing.

- *Blood culture* is the definitive test, which establishes the definitive diagnosis. Alternatively, a bone-marrow culture can also be performed. It correlates well with blood culture in that culture of blood is positive in seventy percent of cases, while bone marrow is positive in ninety percent of cases.

- *Serology* tests, which can detect the presence of antibodies against *Brucella,* are useful. Unfortunately, there are many cross-reacting germs, such as cholera and tularemia bacteria. Hence interpretation caution must be assured. Remember that serologic antibody tests are usually retrospective, since it takes almost two weeks to develop antibodies.

Treatment

The recommended treatment is a combination therapy for six weeks:

1. Doxycycline 200mg per day
        plus
   Rifampin 600–900mg per day

2. Ofloxacin 400mg per day
        plus
   Rifampin 600mg per day

For prophylaxis purposes the doxycycline and rifampin regimen can be used for three weeks.

There are numerous killed and live attenuated human vaccines available in several countries, but they have no proven usefulness. In the past, vaccines have been directed against *B. abortus* or *B. melitensis* for animal use and have proven to be very successful. An effective vaccine against *B. suis* for human use, using killed brucellae, is under development.

PROTECTIVE RESPONSE STRATEGY

- Although *Brucella* species have a long incubation period and the onset of brucellosis is slow, the characteristic *undulant fever* syndrome helps make a diagnosis so that specific treatment can be begun.

- *Brucella* has such a low fatality rate that it would not make a good biological weapon. With appropriate antibiotic therapy, brucellosis should rarely be fatal, regardless of the species used in a bioterror attack.

- Brucellosis cannot be transferred person-to-person, so isolation of patients is not required and infections would be limited to those individuals who inhaled the germs. The exception would be a draining lesion. In this latter case, contact precautions are indicated.

- *Brucella* have no spore form so they can be killed quite readily with any common germicide.

- *Brucella* can be killed by simple pasteurization temperature (145°F for thirty minutes), so if food were contaminated with this germ it could easily be killed.

- Remember that touching your eyes (conjunctiva) is one of the ways to contract brucellosis, so hand washing with soap and water or using a germicidal soap is always an important protective strategy.

- Routine washing of clothing would effectively destroy the bacterium.

6.

TULAREMIA

Bioterror Agent	*Francisella tularensis*
Type of Weapon	Bacterium
Disease Caused	Tularemia
Transmissible (person-to-person)	Yes
Incubation Period	1 to 21 days
Cardinal Features of Disease	Pneumonia without productive cough, lymph nodes swollen at area of entry, in systemic disease no swollen lymph but pneumonia can occur as above
Treatment	Antibiotics: doxycycline, gentamicin, streptomycin
Vaccines Available	Yes

HISTORY AND BACKGROUND

The causative agent of tularemia is *Francisella tularensis*. These bacteria are gram-negative bacilli that are nonmotile and have no spore form. Humans acquire the disease in nature by contact with animals, usually through the inoculation of skin or their mucous membranes with blood or tissue fluids of infected animals or bites from infected ticks, mosquitoes, or flies. A less common method of acquiring infection is through inhalation of contaminated dusts or ingestion of contaminated foods or water. A bioterrorist would likely opt to attack via an aerosol assault, because, as I'll explain later, such delivery of *Francisella* would result in *typhoidal tularemia,* which has a fatality rate of greater than ten percent. There are six forms of tularemia, which I'll discuss in greater detail under clinical features:

1. Ulceroglandular tularemia

2. Glandular tularemia

3. Typhoidal tularemia

4. Oculoglandular tularemia

5. Oropharyngeal tularemia

6. Pneumonic tularemia

Tularemia is sometimes referred to as "rabbit fever." Reservoirs of the bacterium in nature include wild rodents, rabbits, squirrels, muskrats, beavers, voles, deer, and raccoons. Domestic animals can also carry the germ such as cattle, sheep, horses, swine, even cats and dogs. Hence, like many bioterrorist agents, *Francisella* is a zoonotic disease.

The bacterium is highly contagious; as few as twenty-five organisms inhaled, or as few as ten administered subcutaneously (under the skin), can cause infection. The smallest break in the skin can serve as the portal of entry of the germ. In fact, the bacterium is so contagious that numerous cases of tularemia have resulted from laboratory accidents while people were processing infected samples or working with the organisms in research. *Francisella* is distributed worldwide. In the United States, the disease has been reported primarily in southern and south-central states, including Missouri, Kansas, Oklahoma, Arkansas, and Texas.

The first description of the disease tularemia was published in 1911 by George McCoy, who was investigating the bubonic plague outbreak following the devastating San Francisco earthquake of 1906. He eventually named the bacterium *Bacterium tularense*, after Tulare County, California, the site of his laboratory. Dr. Edward Francis, who studied the disease called "deer-fly fever," made the connection between this illness and the plaguelike illness described earlier by Dr. McCoy. For his pioneering work on this disease he was awarded the 1959 Nobel Prize in Science, and the name of the organism was changed to *Francisella tularensis* in his honor.

F. tularensis produces a capsule that allows the organism to avoid immediate destruction by the body's defensive phagocytes. In fact, it can actually survive as an intracellular parasite within the lymph system.

CLINICAL FEATURES

There are six major syndromes associated with tularemia infection, primarily delineated by the mode of organism acquisition.

1. *Ulceroglandular tularemia* is the most common form. It makes up from seventy to eighty-five percent of all cases. A patient would typically present with an ulcer on the skin, which is usually the result of a tick bite. There would also be swollen lymph nodes that drain the site nearest the ulcer. Patients can also experience fever, chills, headache, sweating, and coughing.

2. *Glandular tularemia* makes up about five to twelve percent of cases and is characterized by fever and swollen lymph nodes, but no skin lesion may be apparent.

3. *Typhoidal tularemia* presents with acute onset of fever, chills, headache, vomiting, and diarrhea. It is a systemic disease. Usually no skin

lesion or swollen lymph nodes are seen. Note that the typhoidal type of tularemia is the only form in which diarrhea is usually seen. This form also has the highest mortality rate associated with it, and would likely be a terrorist's choice. About seven to fourteen percent of all natural cases of tularemia are typhoidal.

4. *Oculoglandular tularemia* involves those patients with severe conjunctivitis and swollen lymph nodes as a result of inoculating the organism into the conjunctivae. In nature one to two percent of cases are of this variety.

5. *Oropharyngeal tularemia* occurs in patients who have a primary lesion in the oropharynx. Patients present with severe headache and bilateral tonsillitis or severe streptococcal-type sore throat. Persistent swollen lymph nodes in the neck appear after one to two weeks.

6. *Pneumonic tularemia* occurs in about eight to thirteen percent of all cases, primarily as a complication of any of the other five forms. This type is also acquired by inhalation of infectious aerosols or by dissemination from a bloodstream infection. Lymph nodes in the lungs become swollen.

In any of the tularemia syndromes, whenever the lymph nodes swell they may remain enlarged for long periods of time and eventually they may become necrotic and drain, also for long periods of time. Pneumonia can be associated with any form of the disease but is common in typhoidal tularemia. In all cases there is fever (usually low-grade), malaise, headache, and pain involved in the regional lymph nodes. In a bioterrorist attack with *F. tularensis*, the most likely delivery would be via aerosol dispersal. This would cause primary typhoidal (systemic) tularemia with either a primary or secondary pneumonia. But on occasion the pneumonia may not be evident.

Diagnosis

Making a clinical diagnosis of tularemia can be difficult even though the initial symptoms can be quite severe. This is because the initial symptoms may not be specific enough to allow a diagnosis. A differential diagnosis would require ruling out other typhoidal syndromes like those caused by salmonella, rickettsia, and malaria, or pneumonic processes such as plague or even staphylococcal enterotoxin B disease (to be discussed later). However, as with most bioterrorist attacks, there would be a cluster of individuals presenting with a similar systemic illness. This might provide a clue that a BT attack had occurred. There is another cardinal symptom that might clue in a physician, and that is the lack of a productive cough associated with pneumonia in tularemic patients.

- *Laboratory tests* are generally not helpful. Even routine culture can be difficult because of unusual growth requirements of *Francisella* bacterial species or because of overgrowth with normal flora bacteria that can be grown on special media. Blood cultures may be positive for the growth of a very fastidious bacterium that would eventually be identified. But *F. tularensis* does not grow on ordinary bacteriologic media but would appear in small colonies on special cysteine blood agar at 37°C; however, one must suspect the organism in order to use this special medium. Another problem connected with the identification of *Francisella* is that a specific biologic safety cabinet is required for culturing this bacterium and extreme caution is indicated in order to prevent aerosol acquisition of this germ in the lab.

- *Special tests:* Nasal, sputum, and induced respiratory secretions can be used for PCR and FA analysis by special regional labs. There is a serology blood (serum) test that can be performed called the febrile agglutination antibody test. This might prove useful as a retrospective diagnostic test because it requires a week or more to become positive. There's also a newer ELISA test just made available.

Treatment

For actual therapeutic purposes:

1. Streptomycin (1 gram every 12 hours [IM] for ten to fourteen days)

2. Gentamicin (3–5mg per kilo per day injection for ten to fourteen days)

For chemoprophylaxis purposes the following antibiotics are useful:

1. Doxycycline (100mg twice a day for fourteen days)

2. Tetracycline (2g per day for fourteen days)

The tetracycline drugs have also been used for treatment but are associated with significant relapse rate. Hence they've been relegated to prophylactic regimens only.

There is a live attenuated vaccine for tularemia that is available, but at this time it is only for investigational use. So far more than five thousand people have been vaccinated and none have shown significant adverse reactions, and the vaccine has proven to be efficacious in preventing lab-acquired tularemia. It remains to be tested against a bioterrorist attack, however.

PROTECTIVE RESPONSE STRATEGY

• The single most important thing that anyone can do to protect himself or herself is hand washing. Since tularemia can be transmitted by contact with the organism either via inoculation of skin or mucous membranes, any aerosolized BT attack would leave sufficient levels of bioterrorist agents in the environment where secondary exposure is probable.

• Tularemia bacteria are easy to render harmless by simple application of heat (145°F for thirty minutes), especially since there are no spores associated with this organism, only vegetativ~ forms.

• As for *Brucella*, *Francisella* also has a relatively low fatality rate associated with it, such that it doesn't make a good biological weapon. With adequate antibiotic therapy for ten to fourteen days, there is almost uniform rapid improvement in infected patients.

• Since tularemia can be transmitted to us by contact with household pets like cats and dogs, it is important to observe any changing habits or health status of your pet. They can be the sentinel event to warn you of an impending disaster—much like the "canary in the mine."

- Rarely can the germ be transmitted by the food or water route; nevertheless, being alert to that possibility is important. Cooking food renders the bacterium harmless. Filtering or heating water does the same.

- Routine washing of clothing would effectively destroy the bacterium.

7.

Q FEVER

Bioterror Agent	*Coxiella burnetii*
Type of Weapon	Bacterium
Disease Caused	Q fever
Transmissible (person-to-person)	Rare
Incubation Period	2 to 14 days
Cardinal Features of Disease	Flu-like initially, febrile illness can last up to 2 weeks, acute nondifferentiated febrile illness
Treatment	Antibiotics: doxycycline, erythromycin, rifampin
Vaccines Available	Yes

HISTORY AND BACKGROUND

Q fever is also called query fever because the
causative agent of this disease was initially not
known. The first description of Q fever was made in
Queensland, Australia. The causative agent was
finally identified as *Coxiella burnetii* in 1937. Q fever
is endemic worldwide except in Scandinavia. *C. burnetii* causes an acute systemic infection that primarily affects the lung. The organism is related to rickettsia bacteria (like those that cause Rocky Mountain
spotted fever). Unlike rickettsia, *C. burnetii* can survive extracellularly. Quite unusually this organism
has a sporelike cycle and can exist in two different
antigenic states. If *C. burnetii* is isolated from animals it is in what is referred to as phase I, which is its
highly infectious state. In its phase II form, *C. burnetii* is able to grow in cultural cell lines and is not at
all infectious. Q fever is really a zoonotic disease that
has a natural reservoir in sheep, goats, and cattle.
The bacteria can also be excreted in animal milk,
urine, and feces. Humans acquire the disease by
inhalation of contaminated aerosols. Q fever may
also be transmitted through infected milk or even by
the bite of an infected arthropod. While *C. burnetii*
usually causes a self-limiting febrile illness, a BT
would nevertheless be interested because of its resis-
tance to desiccation and sunlight and its ability to
withstand harsh environmental conditions. Most

important, a *single* bacterial cell can produce clinical illness. That fact ranks *C. burnetii* as a highly infectious agent and gives it a strong likelihood of causing disease if it is aerosolly delivered. This germ would likely be employed as a biological weapon on soldiers before an upcoming battle, perhaps to incapacitate them. As far as its use by a bioterrorist, I don't believe that it would be probable—especially because of its inability to cause critical illness. Infection with *Coxiella burnetii* is more an annoyance than a life-threatening event.

The incubation period varies from two to fourteen days with an average of seven days. Rarely can the period extend to one month. After infection and proliferation in the lungs, the organisms are picked up by the macrophages and carried to the lymph nodes and from there they reach the bloodstream.

CLINICAL FEATURES

After the incubation period, initial clinical manifestations of *C. burnetii* infection include fever, cough, chills, myalgia, headache, and sometimes a pleuritic chest pain. The febrile illness can last from two days to two weeks. The disease generally presents as an acute nondifferentiated febrile illness. Chest X-rays when abnormal (fifty percent of patients) show patchy infiltrates that can resemble viral disease. Some uncommon complications can include endocarditis, hepatitis, aseptic meningitis, encephalitis, and osteomyelitis. Most patients who develop endocarditis have preexisting valvular heart disease.

Diagnosis

Because Q fever usually presents as an undifferentiated febrile illness, similar diseases must be ruled out. Included among diseases that can cause pneumonia are infections that are caused by *Legionella, Mycoplasma, Chlamydia,* and some viral infections. Certainly to be included in the differential should be tularemia and the plague, both of which can mimic Q fever.

- *Culture* of *Coxiella burnetii* can be performed using shell vial cell cultures of human lung fibroblasts, which grow the organism from the buffy coat (phagocytic white blood cells) of blood, as well as from other tissue.

- *Nasal swabs* as well as sputum and induced respiratory secretions can be used for both culture as well as PCR. Blood specimens can be cultured in eggs or mouse inoculation and can also be used for PCR assay.

- *Special tests:* Serologic (blood) tests to identify antibody to *C. burnetii* by IFA (indirect fluorescent) antibody tests or ELISA assays.

Treatment

Various antibiotic agents are useful to treat *Coxiella* infections:

1. Tetracycline 250mg every six hours for five to seven days

2. Doxycycline 100mg every twelve hours for five to seven days

3. Erythromycin 500mg every six hours plus Rifampin 600mg per day for five to seven days

In cases of endocarditis, treatment for twelve months or longer has been successful using:

1. Doxycycline plus rifampin

2. Trimethoprim-sulfamethoxazole plus either doxycycline or tetracycline

A formalized whole cell vaccine is available for investigational use. Q fever vaccine is available and is licensed in Australia. One single dose of this vaccine provides complete protection against naturally occurring Q fever and greater than ninety-five percent protection against aerosol exposure. Because there are some minor reactions to the vaccine, a newer version is under development for use in sensitized persons.

Antibiotics can also be used for prophylaxis. Tetracycline or doxycycline can be given after exposure, which can prevent symptoms or delay the onset of symptoms.

PROTECTIVE RESPONSE STRATEGY

- As previously mentioned, because Q fever is rarely fatal, I believe that it is *not* a likely agent of a bioterrorist. If it is used at all, it would probably be used on soldiers in the battlefield in order to incapacitate them and compromise their fighting performance. Hence, limited response strategies are presented.

- *Coxiella* is only rarely transmitted person-to-person. It is not considered to be a highly contagious disease.

- Hand washing is a very effective means of decontaminating hands. Additionally, the *Coxiella* are easily killed by the action of simple germicides like alcohol and most ordinary germicidal agents.

- Routine washing of clothing would effectively destroy the bacterium.

8.

CHOLERA

Bioterror Agent	*Vibrio cholerae*
Type of Weapon	Bacterium
Disease Caused	Cholera
Transmissible (person-to-person)	Rare
Incubation Period	4 hours to 5 days
Cardinal Features of Disease	"rice water" stools, diarrhea *without* abdominal cramps
Treatment	Antibiotics: doxycycline, ciprofloxacin, erythromycin
Vaccines Available	Yes

HISTORY AND BACKGROUND

The causative agent of cholera is *Vibrio cholerae,* a short, curved, motile, gram-negative non-spore-forming bacillus. Two serogroups have been associated with human cholera infection, namely, 01 and 0139. The 01 group has two other biotypes—the "classical" and El Tor. The bacteria grow best at a neutral pH (7) but are able to tolerate an alkaline condition (pH 9).

Humans acquire the disease by consuming water or food contaminated with the organism. If a bioterrorist were to employ cholera bacteria as a weapon, they would most likely be used to contaminate water supplies. According to military experts, cholera is unlikely to be used in aerosol form.

The organism was first described and named by Pacini in 1854. Later Koch isolated the bacterium that he called "Kommabacillus" because of its characteristic comma or curved shape. Cholera bacteria don't invade the intestinal mucosa, but rather they adhere to it. All strains of V. *cholerae* produce a protein enterotoxin that causes fluid loss in the small intestine but not the large intestine (colon), which is relatively insensitive to that toxin.

There have been seven pandemics (widespread epidemics) of cholera in recorded world history. The first pandemic, in 1816 to 1817, was followed by six others at intervals of about ten to fifteen years. The last occurred in 1961, involving the El Tor strain, whereas all of the others involved the "classical" strain.

In 1989, almost fifty thousand cases of cholera were reported to the World Health Organization from thirty-five countries, which attests to the widespread nature of the current pandemic. Most of these infections have occurred following ingestion of contaminated or poorly cooked seafood or water. Although wound infections have also been reported, these have followed a trauma of some sort, such as swimming or working in infected waters, or exposure to marine animals.

Cholera is endemic in India and Southeast Asia. From these centers cholera has been spread to other countries via shipping lanes. The seventh pandemic of the disease started in 1961 and caused disease in Asia, the Middle East, and Africa. Starting in 1991, the seventh pandemic spread to countries of South America and Central America. Although the disease is rare in North America since the 1800s, an endemic focus exists on the gulf coast of Louisiana and Texas.

The existence of *V. cholerae* outside the human gastrointestinal tract between epidemics and pandemics is uncertain. The bacteria may survive in a dormant state in brackish or salt water. Human carriers are known to exist but are thought to be uncommon.

V. cholerae produces several toxins and other virulence factors, but the cholera toxin (CT) is the most important of these. CT causes the mucosal cells of the small intestine to hypersecrete water and electrolytes into the lumen of the GI tract. This results in a profuse watery diarrhea, which can lead to severe dehydration syndrome. Patients become severely hypotensive, which can lead to death without medical intervention.

The infective dose can be anywhere from ten to five hundred organisms. The incubation period can be as short as four hours or as long as five days but is usually two to three days. The bacteria are easily killed by drying. Interestingly, the bacterium does not survive in pure water but will survive up to twenty-four hours in sewage, and as long as six weeks in water with organic debris in it. It can survive freezing for up to four days. However it is readily killed by dry heat at 225°F or with steam or boiling. It is easily killed by chlorination of water or by short exposure to ordinary disinfectants.

CLINICAL FEATURES

The clinical course begins with sudden onset of nausea and vomiting and a profuse diarrhea *without* abdominal cramps. The stools produced by vibrio infection are characteristically called "rice water"–like, with much mucus, many epithelial cells, and numerous vibrio bacteria. There is a rapid loss of fluid and electrolytes and the resultant dehydration leads to circulatory collapse and kidney shutdown.

The mortality rate without treatment can be as high as fifty percent. In general the El Tor biotype provides a milder clinical course compared with the classical type.

Diagnosis

Can be made by a competent clinician. The character-istic "rice water" stools *without* abdominal cramping hints at a diagnosis of cholera. There is little or no fever associated with this disease and some laboratory characteristics also help with the diagnosis. While there are numerous other pathogens that can cause a watery diarrhea syndrome, such as enterotoxigenic *E. coli*, rotavirus, or other viruses, and ingestion of pre-formed toxins of *Staphylococcus aureus, Clostridium perfringens,* and *Bacillus cereus,* to name a few, the following lab data should help differentiate cholera from these.

- *Microscopic* examination by gram stain will show characteristic curve-shaped gram-negative bacilli. There is also a conspicuous absence of the phagocytes (WBCs) found in stools from diarrhea caused by most other bacterial pathogens. By looking at a drop of stool under the microscope one can also observe the charac-teristic darting motility of vibrio bacteria.

- *Culture* study will reveal typical colonies of *Vibrio cholerae* on a specialized agar media called TCBS (thiosulfate citrate bile salts sucrose) agar. They grow well and produce either yellow or green colonies dependent upon whether they can ferment sucrose (sucrose fermenters produce yellow colonies).

But *Vibrio* also grows on ordinary bacteriological media.

- *Special tests: Vibrio cholerae* can also be identified by using a simple slide agglutination test using specially prepared antisera.

Treatment

Because of the severe dehydration caused by cholera infection, the most important therapy is fluid and electrolyte replacement. Many antibiotics are effective against V. *cholerae.*

1. Tetracycline 500mg every six hours for three days

2. Doxycycline 300mg once
        or
   100mg every twelve hours for three days

3. Ciprofloxacin 500mg every twelve hours for three days

4. Erythromycin 500mg every six hours for three days

There are several other antibiotic choices if need be. For prophylaxis there is a killed vaccine available for use in those considered to be at risk of exposure, but it only provides about fifty percent protection and lasts for no more than six months. There is an inactivated oral vaccine, licensed in Europe, that has shown to be safe, but it also provides short-term protection. Boosters are required every six months. There is no vaccine data against aerosol transmission of *Vibrio cholerae;* however, because the bacterium is unstable in aerosols, it is unlikely that aerosol delivery will be used by a BT.

PROTECTIVE RESPONSE STRATEGY

- Because the major biological threat from this organism appears to be sabotage of food and water, governmental safeguarding of these sources is of paramount importance; so too is the public's vigilance in this regard.

- The likelihood of a BT using cholera bacteria as a biological agent appears to be low, because with appropriate treatment there would be a low death rate as a result of any attack. Hence it is a low-priority agent for use against the general population.

- Even if the agent should be used it would be easily detected in diarrheal feces by ordinary lab culture, in plenty of time to be adequately and successfully treated.

- The bacterium does not survive very well in the environment and is easily killed by simple

germicidal agents. Application of most germicides effectively kills cholera.

- Clinical disease and lab results are very characteristic and aid early diagnosis.

- Rare contact transmission can be easily prevented by good hand-washing techniques.

- Clothing should be washed using a good
germicidal or detergent soap. Underwear
should be washed using bleach.

9.

CLOSTRIDIUM PERFRINGENS TOXINS

Bioterror Agent	*Clostridium perfringens* toxins
Type of Weapon	Toxins of bacteria
Disease Caused	Toxin mediated pulmonary syndrome
Transmissible (person-to-person)	No
Incubation Period	6 hours to 1 day
Cardinal Features of Disease	Acute pulmonary syndrome, leakage of blood vessels and blood vessel destruction
Treatment	Antitoxins
Vaccines Available	No

HISTORY AND BACKGROUND

Clostridia are anaerobic, gram-positive, spore-forming bacilli that morphologically resemble *Bacillus anthracis*. Chapter 4 discussed botulinum toxins produced by *Clostridium botulinum;* this section discusses toxins from another species of *Clostridium* called *perfringens,* which is known to most medical personnel as the organism that causes gas gangrene.

There are many clostridia that can produce gas gangrene (anaerobic destruction of tissue with the production of gas) and myonecrosis (muscle destruction), but *Clostridium perfringens* causes about ninety percent of the cases. This bacterium can produce numerous types of toxins causing a range of medical problems from food poisoning (enterotoxin) to necrotizing enterocolitis (sometimes called "pig bel"), but the most potent toxin of *C. perfringens* is the "alpha toxin," which is really a lecithinase enzyme that splits lecithin, an important constituent of cell membranes. The lethality of alpha toxin is proportionate to the speed at which it splits lecithin into its component parts, phosphorycholine and diglyceride. It is precisely this toxin that a BT would use as a weapon. This alpha toxin would be lethal by aerosol delivery. Other toxins that a terrorist might choose to incorporate into the mix would act to enhance the effectiveness of alpha toxin. Although according to military experts a waterborne delivery of toxin may also be possible, it would be unlikely.

Unknown to many clinicians, *C. perfringens* ranks behind *Salmonella* and *Staphylococcus aureus* as the third most common etiologic agent of food-borne disease in the U.S. Hence if there were a mixed toxin delivered on a population there would be mixed symptoms observed. Strains of *C. perfringens* produce powerful enterotoxins usually when growing in protein (meat) dishes. The enterotoxin produced would cause a hypersecretion to occur in both the jejunum and ileum, with loss of fluids and electrolytes as a patient experiences diarrhea. The precise mechanism has not yet been established. The clostridial toxins produce diarrhea six to eighteen hours after ingestion.

CLINICAL FEATURES

Naturally occurring gas gangrene usually involves rapid invasion, liquefaction (changing solids to liquids), and necrosis of muscles, with gas formation followed by clinical signs of toxicity. The infection spreads in one to three days from an originally contaminated wound site into the subcutaneous tissue and muscle. The result would be a foul-smelling discharge, progressing to necrosis and fever; toxemia, shock, and death might follow in that order. With a naturally occurring food-borne illness, a patient usually develops a crampy abdominal syndrome within six to eighteen hours after eating *C. perfringens*-laden food. The stools produced would be foul-smelling and the diarrhea would have a foamy character. There would occasionally be only vomiting or fever. The toxin forms when the bacteria sporulate in the gut. The entire illness usually lasts one to two days.

Diagnosis

On the other hand, if a BT were to aerosolize *C. perfringens* toxins (especially alpha toxin) as a weapon on a given population, the toxins would produce an acute pulmonary syndrome. Any absorbed alpha toxins (lecithinase) would produce leakage of the blood vessels, hemolysis (blood cell destruction), and liver damage. Of course, the illness would be modified by any one of a number of other toxins that might be included in the aerosol. According to NATO's publicly released document there is insufficient information available to speculate on various clinical syndromes produced by other *C. perfringens* toxins dispersed in an aerosol on humans. They might cause some of the symptoms outlined for the naturally acquired illness caused by *C. perfringens*. However, symptoms might include anemia, elevated liver enzymes, and inability to breathe. Differential diagnosis would make it necessary to rule out other toxin-induced diseases.

- *Culture* would be useless here since only bacterial by-products in the form of toxins would be used in an attack. Various blood (serum) and tissue samples would have to be processed by special reference laboratories of the state, city, or federal government. Nevertheless, if any bacteria were present they could readily be grown under anaerobic conditions on ordinary agar media found in most clinical laboratories.

- *Special tests:* There are numerous PCR and immunoassays available to test for different components of bacteria, including toxins. Certainly even the presence of the toxin enzyme lecithinase can be detected by using a simple egg-yolk medium.

Treatment

There is no specific treatment for *C. perfringens* toxins, even though the organism is susceptible to the antibiotic penicillin. This would in fact be the drug of choice if it were a naturally acquired infection. Recent lab data indicated that the antibiotics clindamycin and rifampin may actually suppress toxin formation. There are polyvalent antitoxins currently available that contain antibodies to several toxins that have been used in treatment, but not enough data exist to prove efficacy.

PROTECTIVE RESPONSE STRATEGY

- Because toxin distributed by aerosol is not dermally active, ordinary clothing may prove effective at trapping toxin, and simple washing of the skin with soap and water should eliminate any surface contamination.

- A simple bleach solution (1:10) inactivates most protein toxins after ten-minute exposure.

- A protective mask, worn properly, is effective against toxin aerosols.

- Adding toxins to reservoirs or lake water would be unlikely to cause human disease because of a dilution factor.

- In theory, open-air exposure to toxin limits its effectiveness, also because of dilution; therefore, toxin is not an extraordinarily effective weapon.

- Activated charcoal tablets might be administered to victim if he or she is conscious and alert.

10.

STAPHYLOCOCCAL ENTEROTOXIN B (SEB) AND TOXIC SHOCK SYNDROME TOXIN-1 (TSST-1)

Bioterror Agent	*Staphylococcus aureus* SEB and TSST-1 toxins
Type of Weapon	Bacterial toxins
Disease Caused	Toxin-mediated systemic disease
Transmissible (person-to-person)	No
Incubation Period	1 to 6 hours
Cardinal Features of Disease	Fever (102°–104°F), low blood pressure, rash, desquamation, vomiting, nausea, diarrhea, acute respiratory failure, septic shock
Treatment	Antitoxins, steroids
Vaccines Available	Yes (Experimental)

HISTORY AND BACKGROUND

By far *Staphylococcus aureus* is the most important gram-positive human pathogen. It is an organism that plays two roles—sort of like Dr. Jekyll and Mr. Hyde. On one hand it is part of our normal body flora, a bacterium that resides in various body sites, such as the nose, skin, and vagina, without causing harm. On the other hand, it can cause a variety of infections, from superficial skin, wound, respiratory, heart, urinary tract, and blood infections. In addition to infections, *S. aureus* is able to cause toxemias such as food poisoning and toxic shock syndrome. It does so by producing powerful protein exotoxins called enterotoxins.

Staphylococcal food poisoning is one of the most common food-borne illnesses, which is due to toxin-contaminated food rather than an infection caused by the bacterium. Therefore, this type of food-borne illness is called an *intoxication* rather than an infection. Most of us are familiar with the term "picnic diarrhea," which is commonly used to describe this illness. The usual scenario involves consuming foods such as ham, meat, pies (especially custard), salads made with mayonnaise, and cream dishes that have been left out during a picnic or at ambient temperature in a home for a few hours, during which time the staph grow luxuriantly and produce toxins in the process. *S. aureus* is a hearty organism that is very salt-tolerant and can grow well even in ten- to fifteen-percent salt solutions. That's why it can grow well on salt ham or pork. *S. aureus* is the first pathogen discussed thus far that is not a zoonotic germ, and, in fact, its main reservoir is human.

There are at least five enterotoxins produced by *S. aureus* (A to E) in addition to numerous other types of toxins. In fact, it can be said that *S. aureus* is one of the most toxin-producing germs known to man. The enterotoxins are quite resistant to stomach acids and can even survive boiling for thirty minutes. Hence, any food contaminated with enterotoxins will not be made safe by heat. Various studies show that about thirty to fifty percent of *S. aureus* strains are able to produce enterotoxins. The most commonly associated enterotoxin, which causes disease, is Enterotoxin A. Enterotoxins C and D are associated with milk and milk products. Enterotoxin B is a cause of pseudomembranous enterocolitis, and it also causes a type of toxic shock syndrome (TSST-1 is the primary toxin associated with menstrual toxic shock syndrome) that is not menstrually associated—and, of course, it also can cause food poisoning.

The enterotoxins of *S. aureus* cause profuse diarrhea and projectile vomiting because the toxins work directly in the vomiting control center of the brain.

Both *S. aureus* and its enterotoxins are hardy and can survive under adverse conditions. *Staphylococci* are generally arranged in pairs and clusters much like grape clusters. In fact, the Greek word *staphyle* means "a bunch of grapes."

After contaminated food is eaten the onset of symptoms is quite fast. The incubation period is only about four hours. This period is so short because the toxins ingested are preformed. If a limited amount of toxin is ingested, the symptoms will be short-lived, usually less than twenty-four hours. Food poisoning is characterized by abdominal pain, nausea, and vomiting with nonbloody diarrhea. Fever is not usually seen. Severe dehydration can occur, which must be treated with fluid replacement. If a bioterrorist were to use staphylococcal enterotoxin B (SEB) as a weapon, not only would the symptoms be like those of food-borne illness but there would be other more severe symptoms with significant morbidity and mortality. The reason for this is because a BT would dispense SEB in aerosol form, just as one would predict he would do for any toxin weapon. In that case, the symptoms would be significantly different from those of food poisoning. Since such a BT attack has never occurred, we really are only speculating on such symptoms. However, we do have a model that we can refer to. That model is the one for toxic shock syndrome. The toxic shock syndrome toxin-1 is closely related to SEB and in fact, can also be used as a BT weapon in place of SEB. They would give similar symptoms if used as a biological weapon. The following is a brief discussion of TSS and the effects of TSST-1.

Toxic shock syndrome (TSS) is a multisystem disease, meaning it can affect more than one organ system in the body. The symptoms mimic those of scarlet fever and range from a mild, flu-like illness to heart, kidney, liver, and respiratory failure, including fever equal to or greater than 102 degrees Fahrenheit, skin eruptions with red rash and subsequent peeling of the skin, nausea, vomiting, diarrhea, low blood pressure, and dizziness. The disease is caused by exotoxins produced by strains of *Staphylococcus aureus*, which many people have as part of their normal flora but which can prove fatal if its toxin gets into the bloodstream through a wound infection, a complication of influenza, or some other means.

In 1928, a pediatrician in the small city of Bundaberg in Queensland, Australia, inoculated twenty-one children, including his own son, with a vial of diphtheria antitoxin contaminated with *S. aureus*. Within twenty-four hours, eighteen of the twenty-one children became ill, and eleven children died. A twelfth child, the final fatality, died at the beginning of the next day. A royal commission investigated the disaster and reported that patients displayed scarlet-fever-like symptoms. The nine children who survived all developed abscesses where they had been injected with the antitoxin, and these abscesses were found to contain *S. aureus*. In fact, as early as 1908, the medical and scientific literature reported that numerous strains of *S. aureus* caused an illness resembling scarlet fever.

Toxic shock syndrome burst into public attention in 1980, and I became actively involved in research into it at that time. By 1983 more than 2,200 cases had been reported to the Centers for Disease Control in Atlanta. A large proportion of these cases were fatal and followed the same pattern. Typical cases involved young, previously healthy women who were menstruating and using superabsorbent tampons.

Beginning over fifty years ago, microbiologists found that staphylococci require the presence of agar or other thickening agents to produce toxins of enough quantity and potency to cause illness. The thicker, or more viscous, the medium in which S. aureus is growing, the more toxin will be released. This is why staphylococci are likeliest to cause food poisoning in rich, thick dishes like potato, macaroni, and egg salad. The menstrual flow has a relatively high viscosity because of the proteins in the blood. My early work with the 1980 superabsorbent tampons and a glass of water suggested that in the case of women wearing synthetic tampons, the synthetic fibers in the tampons might dramatically increase viscosity in the vagina because of the great amount of blood they would absorb and because of the greater viscosity that synthetic fibers have to begin with.

A second important physical factor is surface area. The more extensive the surface area that S. aureus has available to grow on, the more toxins it will create. By the nature of their materials, synthetic tampons have much more surface area than all-cotton tampons. There are numerous other physical and chemical factors that are provided by using a tampon during menstruation that help to create ideal conditions for toxin production.

To sum up, synthetic tampons, with their superabsorbent materials and greater surface area, create a microenvironment within the vagina that constitutes an ideal breeding ground for any toxin-producing *S. aureus* that a woman may have in her normal vaginal flora or harbor there transiently. It is as if a woman decided to give the toxin-producing *S. aureus* a fortified garrison, in the form of the synthetic tampon, in which to build up its strength and numbers before flooding the bloodstream with toxins. This effectively turns a mutualist germ into a viciously parasitic one.

This also explains one of the most frightening aspects of toxic shock syndrome, which is that surviving it does not confer immunity as would be usual in most diseases. The reason is that the TSS toxins suppress the body's lymphocytes and keep the B cells from transforming into plasma cells that can make antibodies to fight the toxins. The situation is even more dire in tampon-related toxic shock syndrome: The tampon is such a fertile breeding ground for *S. aureus* that it leads to toxin production far greater than would be found in any naturally occurring incidence of the disease.

It is possible to acquire sufficient antibodies against TSS toxins. But this can only happen incrementally, in small steps. As we grow and mature and become exposed to *S. aureus,* our immune systems may have many occasions to respond to minuscule quantities of TSST-1 and other TSS toxins. Over time, the antibodies can become numerous enough to safeguard a person against TSS.

CLINICAL FEATURES

The clinical features of a BT attack using staph enterotoxin B or TSST-1 would be remarkably similar to a person getting TSS as described above. A BT would probably aerosolly deliver the toxins, which would cause a very rapid onset of symptoms from about one to six hours after exposure. These staph toxins are called pyrogens, meaning that they will induce fever. Additional initial symptoms would include headache, myalgia, and nonproductive cough. There may be the appearance of a sunburn-like rash which after ten to fourteen days desquamates (skin peels). Temperature is generally less than 102°F and may reach 106°F. Fever may last several days and depends upon the amount of toxin inhaled. Patients will also likely display nausea, vomiting, and diarrhea. In severe cases, marked pulmonary edema (fluid accumulating in the lungs) may occur, or even what is called ARDS (acute respiratory distress syndrome), which can lead to respiratory failure. If high levels of toxin are ingested, postural (positional) hypotension (low blood pressure) can occur, leading to septic shock and death.

Diagnosis

Certainly there are numerous viruses, such as influenza and adenovirus, as well as bacteria, such as mycoplasma, that may present an initial diagnostic dilemma; however, unlike those diseases most of the symptoms caused by inhalation of SEB or TSST-1 toxin would occur within a relatively short period of time (usually within a twenty-four-hour period). In fact, clinically a physician would have to differentiate from other similar diseases, including tularemia, plague, Q fever, and botulinum intoxication. However, a good clinician can make a clinical diagnosis quite confidently.

Laboratory findings are not very helpful, although there is an increase in the number of white blood cells (WBCs) in particular neutrophils (phagocytes), indicating a possible infection. The body seems to react to toxins as it would to germs. However, a better test to detect toxins might be urine samples, which can be tested by PCR methods or by special toxin assays. In addition to urine, induced respiratory secretions and even nasal swabs may be suitable specimens to assay for the presence of toxins. Patients who are exposed to small amounts of toxin may develop protective antibodies. There are no other test procedures that are useful to help diagnose SEB or TSST-1. Remember that neither is contagious.

Treatment

Intoxications are similar to being poisoned. If an individual is exposed to a specific amount of toxin, that individual either survives the challenge of poison or succumbs to it. Luckily the majority of patients will probably recover from the challenge. Because patients generally become exposed to small amounts of SEB toxin expressed by strains of S. aureus that they may transiently and periodically become exposed to during the course of living their lives, many individuals already have some protective antibodies. The same circumstance applies to TSST-1. However, if the toxin challenge is very high, their protective antibodies might be overwhelmed. Patients who are exposed to toxin are often treated with steroids in order to mollify their symptoms. Further treatment would be related to supportive care, especially compensating for oxygen and fluid losses (which can be quite extensive in cases of SEB or TSST-1 intoxication). While there are currently no vaccines available for either staph toxins, there are experimental vaccines in development that have shown promise. There is data that confirms the usefulness of the vaccine in monkeys. There is even an experimental antitoxin that can be used to reduce mortality if it is given early enough.

PROTECTIVE RESPONSE STRATEGY

- Toxins are not very effective weapons. Open-air exposure limits effectiveness because of a dilution factor. Such toxins may more likely be used in a military situation because they can incapacitate troops.

- SEB toxin and TSST-1 are expected to have a low mortality rate (less than one percent), and as such are not an effective weapon for use on the general public.

- As with other toxins, these staph toxins are not dermally active, and ordinary clothing may also prove effective at trapping toxins. Simple washing of the skin with soap and water should eliminate surface contamination. Routine washing of clothing would remove toxin.

- A ten percent bleach solution inactivates these two protein toxins after a ten-minute exposure.

- For ingested toxins, activated charcoal tablets might be administered to victim if he or she is conscious and alert.

- A protective mask, worn properly, is effective against toxin aerosols.

11.

MELIOIDOSIS AND GLANDERS

Bioterror Agent	*Burkholderia mallei* and *pseudomallei*
Type of Weapon	Bacteria
Disease Caused	Melioidosis and glanders
Transmissible (person-to-person)	Not likely
Incubation Period	3 to 14 days
Cardinal Features of Disease	Pneumonia with fluid around lungs, septicemia, hypotension, and shock
Treatment	Antibiotics: doxycycline, sulfa drugs
Vaccines Available	No

HISTORY AND BACKGROUND

Both melioidosis and glanders are caused by members of the genus *Burkholderia* (formerly genus *Pseudomonas*). *B. mallei* is an obligate gram-negative bacillus and a parasite of animals (primarily horses, but also sheep, goats, mules, and donkeys), and causes a respiratory tract infection known as glanders. *B. mallei* is not indigenous to soil, water, or plants, only animals.

It rarely causes human infection, which occurs in two ways: through an abrasion of the skin and via the respiratory tract. Although this germ mainly infects horses, when man is infected it can be serious. Untreated systemic glanders is almost one hundred percent fatal.

B. pseudomallei is very much related to *B. mallei*. It causes a glanders-like disease, melioidosis, in animals as well as humans. This gram-negative organism is prevalent in Southeast Asia but has been described elsewhere. Melioidosis usually causes disease in rodents but can be transmitted to man via food contaminated by rodent droppings and by biting flies. Lastly it may be transmitted by aerosolization. The mortality rate is ninety-five percent in acute-disease patients who are not treated.

Certainly the primary way a bioterrorist would weaponize both glanders and melioidosis would be to make it an aerosol.

A SHORT HISTORY OF GLANDERS *(B. MALLEI)*

- sporadic cases occur in Africa, Asia, the Middle East, and South America

- During WWI, Russian horses and mules were deliberately infected by Central powers

- post WWI—many Russians had infections

- during WWII, Japanese infected horses and people in China

- U.S. studied weaponization of glanders during WWII

A SHORT HISTORY OF MELIOIDOSIS *(B. PSEUDOMALLEI)*

- exists in soil and stagnant water in an area latitude 20 degrees north and south of the equator (mostly in Thailand and Vietnam)

- estimated that 250,000 U.S. military personnel were probably infected during the war in Vietnam from 1965 to 1973

CLINICAL FEATURES

Glanders

Incubation period ranges from three to fourteen days dependent upon the size of the particles aerosolized and the dose. There are two main forms of glanders, an acute form and a chronic form. The *acute* form is the inhaled form, which affects the upper respiratory tract. In nature, the acute form kills infected animals in three to four weeks. The acute form of infection of the nasal, oral, or conjunctival mucous membranes causes blood-streaked discharge from the nose with nodules and ulcerations. The *chronic* form affects the joints and lymph as multiple skin nodules begin to ulcerate with pus forming within them. Other symptoms include fever, sweats, myalgia, headache, enlarged spleen, and chest pain. Sometimes pneumonia is present. As few as one to ten bacteria aerosolly delivered to animals is lethal!

Melioidosis

This disease is very similar to glanders. Clinical syndrome in humans is characterized by pneumonia with fluid around the lungs. The pneumonia may be the result of inhaling the bacterium during an assault or via being spread by a bloodstream infection. Any infection with this bacterium may result in a septicemia (blood infection), which may, in turn, cause hypotension and shock. Three forms of the disease are recognized:

- *acute* disease, which primarily presents as a bloodstream infection (septicemia)

- *subacute* disease, which very much mimics a tuberculosis-like picture

- *chronic* disease, presents more like a localized cellulitis (inflammation within skin tissue)

Diagnosis

The differential diagnosis for *melioidosis* would include tuberculosis, as well as diseases that cause draining pus in tissue. In fact, the chest X-ray may mimic tuberculosis. The differential diagnosis for glanders would be similar to melioidosis.

Laboratory: B. *mallei* grows slowly but does grow on enriched bacteriological media. Gram-stain smears will reveal small gram-negative bacteria. The culture may take two days for growth. There are several blood tests used to detect antibodies, but these would take up to fourteen days to develop, so they can be used only retrospectively.

B. *pseudomallei* grows well in standard bacteriological media and can be identified by biochemical tests. A gram stain of sputum or wound exudates will show small, gram-negative bipolar-staining bacilli.

Treatment

For glanders: sulfadiazine, doxycycline, rifampin, trimethoprim-sulfamethoxazole, streptomycin, ciprofloxacin for three weeks of therapy. For melioidosis: tetracycline, chloramphenicol, trimethoprim-sulfamethoxazole, doxycycline, and ceftazidime for 60 to 150 days of therapy. For glanders: There is currently no vaccine available for use in humans. Some researchers recommend trimethoprim-sulfamethoxazole for prophylaxis. For melioidosis: There is no vaccine available for use in humans, nor is there a prophylactic antibiotic regime for post-exposure.

Many laboratory workers have been infected by handling the organism carelessly. The following vignette emphasizes three points:

1. The highly infectious nature of the organisms in high concentration.

2. The difficulty of diagnosing the illness yet the ample window of opportunity to make a successful diagnosis.

3. The successful treatment of an infection.

A real-life incident at the U.S. Army's Medical Research Institute of Infectious Diseases at Fort Detrick, Maryland, illustrates how difficult diagnosis can be. In March 2000, a thirty-three-year-old civilian microbiologist working in one of Fort Detrick's high-security labs became ill with fever. The fever started at 100.5°F, then rose quickly to 103°F, accompanied by general malaise and some weight loss over the course of a few days. The researcher went to his private physician, who treated him with antibiotics, and he continued to go to work. But he didn't get better. Eventually, he had to be hospitalized, and finally with the help of U.S. Army doctors he was diagnosed with glanders, a disease of horses that is fatal to human beings if untreated and that has been the subject of weapons experiments by the

United States, the Soviet Union, and other countries. The scientist had been working with eighteen other researchers under rigorous safety conditions in a tightly sealed laboratory with elaborate filtration systems. He had so much confidence in these safety measures that he didn't consider the possibility that something in the lab had made him sick, especially when none of his coworkers became ill. Luckily for him an accurate diagnosis was reached in time, he was treated properly, and he recovered. But if people who work with biological agents every day can be fooled, think how easy it would be for ordinary emergency-room staff and others to interpret the result of a bioterrorist attack as a simple case of the flu or some other common malady.

PROTECTIVE RESPONSE STRATEGY

- Person-to-person transmission is not likely.

- Environmental decontamination can be effective with use of a simple ten percent bleach solution or other ordinary disinfectants like phenolics (one percent).

- Both *Burkholderia* species are killed by heating to $165°$ F.

- The ultraviolet rays of sunlight can kill *Burkholderia* in several hours.

- Neither agent is a good candidate for bioterrorism, because they are detectable by ordinary bacteriological media and are treatable with antibiotics. Therefore that would reduce the high death rate to near zero. These organisms would probably make better military weapons.

12.

VENEZUELAN EQUINE ENCEPHALITIS (VEE)

Bioterror Agent	VEE virus
Type of Weapon	Virus
Disease Caused	Venezuelan equine encephalitis
Transmissible (person-to-person)	Moderate
Incubation Period	4 to 21 days
Cardinal Features of Disease	Flu-like then severe lethargy for up to 2 weeks, then neurologic manifestations in 5%
Treatment	Relieve symptoms
Vaccines Available	Yes

HISTORY AND BACKGROUND

Unquestionably there are many types of encephalitis viruses. VEE is, in fact, clinically (based on symptoms presented) indistinguishable from others such as St. Louis encephalitis, eastern and western equine encephalitis, Japanese B-type encephalitis, Russian Far East encephalitis, and even West Nile encephalitis viruses. The first question that should come to mind is why aren't the others used as a potential BT weapon? Certainly they can be, but the attack rate (that is, the number of people who would probably get the disease after being exposed to the agent) is much lower than that for VEE. VEE has an attack rate of one hundred percent. Can't get better than that. That is precisely the reason the U.S. government weaponized it in the 1950s and 1960s before the U.S. offensive bio-warfare program was terminated. However, it is still considered to be a potential weapon in the hands of many other nations.

VEE can be weaponized either as a liquid or dry form for aerosol dispersal. It can be transmitted in three ways:

1. via mosquitoes: naturally occurring incidence of VEE is low

2. via aerosol, either liquid or dry form

3. secondary spread by person to person is

thought to occur but has not been conclusively shown

In nature VEE is a mosquito-borne viral disease that is neurotrophic (has predilection for the central nervous system) causing encephalitis (inflammation of the brain and the meninges of the brain) in equine animals (horses and mules) and an unremarkable febrile illness in humans. More than fifty percent of equines that become infected develop encephalitis, while in humans almost one hundred percent of those exposed will develop an influenza-like illness! Only two to four percent of patients during a nature epidemic develop signs of central nervous system involvement, with a fatality rate of less than one percent.

In the U.S., VEE is a rare disease. The disease was first reported in Venezuela in 1936. VEE is prevalent in South and Central America, Trinidad, Mexico, and Panama. In 1971 there was a report of a severe epidemic that occurred along the Texas-Mexico border and killed several thousand horses. There were a few hundred human cases also reported during that epidemic. Florida had been another hot spot of VEE, but the U.S. has controlled the disease with an extensive vaccination program of equines, using an attenuated live-virus vaccine. In addition, extensive spraying of mosquitoes was begun sometime after the 1971 outbreak. Routine spraying has continued since. In the case of the West Nile encephalitis virus, a sentinel animal (birds) indicated the presence of disease before human infection occurred. In the case of VEE, that would have been horses, but because we vaccinate them there is no sentinel animal system to warn us that a VEE virus attack has occurred. On the other hand, because we have eradicated VEE from the U.S., any human with the disease would likely signal a BT event.

VEE is an arthropod-borne alphavirus that has been incidentally associated with human disease. Eight serologically distinct viruses exist, but only two are important pathogens for humans: variants A/B and C.

Most encephalitis viruses are destroyed by heat and are easily killed by ordinary disinfectants.

CLINICAL FEATURES

The incubation period ranges from one to five days. Rapid onset of fever (usually high), headache, dizziness, lethargy, depression, anorexia, chills, myalgia, photophobia, nausea, vomiting, cough, sore throat, and even diarrhea may occur. VEE is not distinguishable from other viruses that cause encephalitis. The acute phase of the disease exists from one to three days followed by a prolonged period (up to two weeks) of lethargy. Full recovery is usual after two weeks. It is estimated that inoculation with about ten to one hundred viruses can cause infection. In less than five percent of patients during a natural epidemic would there be any neurologic manifestation, which would be characterized by convulsions, coma, and paralysis. The neurologic cases are seen mainly in children. There's up to a twenty percent death rate possible in children.

Diagnosis

At its onset VEE may be very difficult to differentiate clinically from influenza. Clues may be gleaned only if a cluster of patients come down with similar symptoms and if encephalitis is diagnosed among this population. If there is a neurological component to the illness, VEE cannot be differentiated clinically from other encephalitis viruses.

- *Laboratory* studies will reveal a marked decrease in white blood cells, particularly lymphocytes.

- *Special tests:* It is possible to grow the virus in cell cultures or in suckling mice. There are also special virus-detection ELISA assays that can be performed to detect VEE. Nasal swabs and induced respiratory secretions can be used for both culture and PCR. Blood (serum) can also be used for culture. In addition, there are a large number of blood serology tests that can be performed in order to detect antibodies to VEE using methods such as ELISA, fluorescent antibody, and hemagglutination inhibition tests, among others.

Treatment

There is no specific antiviral therapy for VEE. The treatment would be geared toward relieving symptoms of headache or myalgia, controlling the convulsions that sometimes occur, or aiding any difficulty in breathing. Although there are several new experimental drugs that have shown some promise in the treatment and prophylaxis of VEE, there is insufficient human data. An experimental live, attenuated vaccine called TC-83 has shown some efficacy. In tests using several thousand people, it has prevented laboratory infections. The TC-83 vaccine is licensed for use in equine animals. There is a second experimental vaccine, designated TC-84, that is currently being tested. The vaccines are given in a single subcutaneous (below the skin) dose. Both have a nonresponder rate of about twenty percent.

PROTECTIVE RESPONSE STRATEGY

- VEE has a very low lethality and in the majority of victims it might inflict only a flu-like disease. I believe this weapon would be of low priority.

- If an outbreak did occur, control of the mosquito population would be mandatory. A similar program as that employed for West Nile virus would be required.

- Rare person-to-person transmission (via contact spread) can be prevented by the practice of good hand-washing techniques.

- The VEE virus is easily destroyed by simple germicides such as ten percent bleach and Lysol, as well as heat ($165°F$).

- Contaminated clothing can be washed clean with any detergent soap.

13.

CRIMEAN CONGO HEMORRHAGIC FEVER AND OTHER VIRAL HEMORRHAGIC FEVERS

Bioterror Agent	Crimean Congo and other viral hemorrhagic fevers
Type of Weapon	Virus
Disease Caused	Hemorrhagic fevers
Transmissible (person-to-person)	Some
Incubation Period	4 to 21 days
Cardinal Features of Disease	Vascular damage, petechiae, high fevers, bleeding from orifices
Treatment	Treat symptoms; immune globulins
Vaccines Available	Yes, for some

HISTORY AND BACKGROUND

Crimean Congo hemorrhagic fever (CCHF) is only one of many illnesses referred to as viral hemorrhagic fevers (VHFs). Among these are Ebola and Marburg viruses of the family Filoviridae; Lassa fever, Argentine (Junin) and Bolivian hemorrhagic fevers of the family Arenaviridae; Crimean Congo hemorrhagic fever, hantavirus, Rift Valley fever of the family Phlebovirus; and dengue hemorrhagic fever and yellow fever virus of the Flaviviridae family. Any of these hemorrhagic viruses can be weaponized via an aerosol delivery.

Crimean Congo hemorrhagic fever is transmitted primarily by ticks and occurs in Africa, the Balkans, Europe, Asia, the Middle East, and the former USSR. Dengue is a mosquito-borne infection endemic in more than one hundred countries in Africa, North and South America, the Mediterranean area, Asia, and the Western Pacific region. The Ebola virus is transmitted via direct contact with blood, secretions, organs, or semen of infected patients. The Marburg virus is closely related to Ebola and was first identified in Marburg, Germany (where it got its name), when African green monkeys were shipped from Africa to Germany. This virus is indigenous to sub-Saharan Africa. It can be spread by direct contact with infected blood, secretions, organs, and semen in much the same way Ebola is spread. Rift Valley fever virus occurs only in sub-Saharan Africa and is mosquito-borne but can also be spread by aerosols. Argentine hemorrhagic fever, caused by Junin virus, was first identified in 1955 and is spread in nature by contact with infected rodent droppings. Bolivian hemorrhagic fever, caused by Machupo virus, was first described in northeastern Bolivia. Both fevers are also spread by dried rodent excreta via aerosol transmission. Lassa fever virus is related to both the Argentine and Bolivian hemorrhagic fever viruses but is distributed in Western Africa. Hantavirus was described before World War II in Manchuria, but hemorrhagic disease has been reported in Europe as well. Yellow fever is naturally transmitted by the bite of a female mosquito. Both yellow fever and dengue fever can cause a hemorrhagic fever syndrome. Interestingly, all of the aforementioned viruses except dengue can also be infectious via aerosols.

Crimean Congo hemorrhagic fever is transmitted by ticks or by the crushing of infected ticks or by the slaughter of infected livestock. Several reports describe spread of CCHF in hospitals, where person-to-person spread was shown to be possible. As few as one to ten viral particles can cause infection.

CLINICAL FEATURES

Even though I am highlighting CCHF, the general clinical syndrome associated with all of the afore-

mentioned viruses is similar and is called "viral hemorrhagic fever" or VHF. Although there are many
variables, including host responses and viral strain
differences, that would give rise to variation in specific symptoms recorded, I am taking the liberty of
generalizing, for the sake of convenience and sim-
plicity.

The most common presenting symptoms are fever,
myalgia, low blood pressure, flushing and ecchymoses
(black-and-blue marks) anywhere on the body.
Typically the onset of CCHF is three to twelve days
after tick exposure or inhalation of an aerosol. There
can be extensive GI bleeding and extensive ecchymoses. Among other symptoms experienced are
headache, back pain, nausea, vomiting, and delirium.
A jaundiced appearance (a yellow coloration of skin
and whites of the eyes, which is evidence of liver failure) is also seen, as is a swollen liver (hepatomegaly).
Mortality for CCHF is fifteen to thirty percent, but
some hemorrhagic fevers, such as Ebola can have a
death rate near ninety percent.

Diagnosis

The key to diagnosis is a vascular involvement (evidence of blood vessel damage), which would be evidenced by petechiae (pinpoint-sized hemorrhagic
spots under the skin), easy bleeding, flushing of the
face, postural (related to changing position) hypotension, and edema. We must keep in mind that the *tar-*

get organ in VHF is the vascular system; hence the aforementioned symptoms relate to microvascular damage and changes in permeability (movement of fluid out of blood vessels into surrounding tissue, giving rise to edema). In addition, high fever, dizziness, and GI bleeding may be present. Differential diagnosis would have to be made by an experienced clinician to identify the particular VHF causing the illness as well as differentiating a few other diseases that can mimic VHF in its early stages, such as rickettsial diseases, meningitis, and some others. Fatal cases of VHF are associated with extensive hemorrhage (bleeding), even from needle-puncture sites; coma; and shock.

- *Laboratory* findings would clearly indicate leukopenia (low number of white blood cells) and thrombocytopenia (low numbers of platelets that are responsible for normal clotting of blood). In addition, a special liver enzyme called AST (aspartate aminotransferase) would become elevated, indicating liver damage.

- *Special tests:* There are numerous rapid enzyme-immunoassays (ELISA and EIA) available to detect the different VHFs. There are also many types of blood serology tests (FA and ELISA) that can be used to detect antibodies to the VHF diseases. Virus isolation is also possible by injecting blood from acutely ill patients into suckling mice.

Treatment

The most important and immediate care needed is the management of hypotension caused by fluid loss. Very aggressive supportive care must be provided. There aren't many specific drugs that can be used against the myriad of VHFs, but the antiviral drug ribavirin can be given to CCHF patients. Treatment with immune globulin vaccines have shown usefulness against CCHF, RVF, BHF, and Lassa fever. There are several vaccines under development:

Argentine HF vaccine (live attenuated)

Bolivian HF vaccine (same vaccine)

Rift VF vaccine (both live and inactivated)

Yellow fever vaccine is the only established and licensed vaccine for any of the hemorrhagic fevers.

For prophylaxis in the event of a suspected or known biological attack, ribavirin would probably be used.

PROTECTIVE RESPONSE STRATEGY

- Any individual who may have been exposed to blood, body fluids, secretions, or
  excretions from a patient with a suspected
  VHF should immediately and thoroughly
  wash the skin surfaces with soap and
  water. Preferably a complete shower should
  be taken. The following vignette exemp-
  lifies the ease of transfer of Ebola hemorrhagic fever and emphasizes the need for
  caution:

In the fall of 2000, newspapers across the world
announced an outbreak of Ebola fever in the
Ugandan district of Gulu. Ebola is another disease
that passed to humans from among our close rela-
tives, this time from a certain monkey species. It can
be transmitted from person to person by sexual contact. But as we've already seen, this terrifying disease, which literally liquefies a person's insides,
causing profuse internal and external bleeding, can
also be passed by touch or even respiratory secretions. In Uganda this easy transmission has been
enhanced by traditional funeral practices, in which
family members and close friends bathe the
deceased person and then wash their hands in a
communal bowl as a sign of unity. This practice trig-

gered the recent outbreak in Uganda, when a thirty-six-year-old woman, the mother of two children, fell ill and died in her mud hut. Not knowing that she had died of Ebola fever, the woman's mother, her three sisters, one of her children, and three of her friends gathered together to bathe her body as a sign of their respect and love. All of them have since died of Ebola fever. In total this particular outbreak took more than ninety lives, including those of some medical personnel, before it was contained by procedures for isolating the dead and insuring that caregivers did not contaminate themselves by failing to wear goggles, mask, and gloves. A complicating factor here is that in many underdeveloped countries basic medical supplies that we take for granted in the West may be in short supply or unavailable. And in the midst of a crisis, overworked doctors and nurses can easily lose sight of protecting themselves as they try to help others.

While the Ebola episode described above is not applicable to all of the VHFs, it does emphasize the need to prevent contamination of oneself and the environment. Remember there are many unknowns regarding any BT event! Too much caution can never be enough!

- In hospitals, contact precautions should be instituted to prevent contagion. A special respiratory mask called N95 is recommended and should be worn for protection. Infected patients should be quarantined for the duration of the illness.

- Anyone caring for (or visiting during convalescence) a patient that is stricken with a VHF must practice strict "barrier nursing" techniques (in other words, completely protect oneself from the infected patient by dressing in gown, gloves, mask, eye protection, hat, booties, etc.), because evidence has accumulated indicating that large droplets or even fomites (inanimate objects) may act as mediators of transmission of the virus.

- Any contamination of mucous membranes must be immediately diluted with copious quantities of water. Contamination of eyes can be diluted with saline or OTC products like Visine.

- Decontamination of surfaces, equipment, or other articles can be accomplished by using a ten percent solution of Clorox. Lysol-type disinfectants are an alternative.

- Deceased individuals should be sealed in leakproof material for prompt burial or cremation.

- VHFs are very difficult for a bioterrorist to weaponize, and there is no real carrier state for VHF, making it an unlikely weapon.

- Articles of clothing should either be discarded if heavily soiled, or washed well using a strong detergent containing a germicide, or washed using bleach.

14.

RIFT VALLEY FEVER

Bioterror Agent	Rift Valley fever virus (RVF)
Type of Weapon	Virus
Disease Caused	Hemorrhagic fever
Transmissible (person-to-person)	Yes
Incubation Period	4 to 21 days
Cardinal Features of Disease	Vascular damage, petechiae, high fevers, bleeding from orifices
Treatment	Treat symptoms; immune globulin
Vaccines Available	Yes

HISTORY AND BACKGROUND

Although Rift Valley fever (RVF) is one of many VHFs described in the preceding section, it deserves special mention just as Crimean Congo hemorrhagic fever did in the previous section for a number of reasons: it is transmitted by mosquitoes as well as by aerosols, there is an inactivated vaccine available for prevention, and it appears to be sensitive to the antiviral drug ribavirin. These features are useful and make the study of RVF serve as a model or standard for the study of other VHFs. The following chart lists the VHFs presented in Chapters 13 and 14 and their natural means of transmission:

VHF Agent	Natural Means of Transmission
Ebola	Contact
Marburg	Contact
Lassa fever	Contact
Argentine (Junin)	Contact and aerosol
Bolivian (Machupo)	Contact and aerosol
Crimean Congo	Ticks and contact
Hantavirus	Contact and aerosol
Rift Valley fever	Mosquito and aerosol
Dengue	Mosquito
Yellow fever	Mosquito

Rather than repeating the details outlined in Chapter 13, a few highlights of RVF are presented for comparative study purposes.

RVF disease occurs primarily in sub-Saharan Africa, transmitted by mosquitoes. In addition to being infected by biting arthropods, it is also possible for humans to become infected by aerosols. This is of particular importance, because if a bioterrorist were to use RVF as an attack weapon he would aerosolize it for delivery. In nature, outbreaks of RVF are associated with a heavy rainy season with a subsequently heavy population of mosquitoes. During an attack by a bioterrorist, domesticated animals can become infected by aerosol means; after infection, the resident mosquito population can continue the assault. Hence RVF can teach us a lesson regarding development of an effective defensive strategy that must also include mosquito-control techniques.

CLINICAL FEATURES

The incubation period ranges from two to five days and as the name of the disease implies is hallmarked by a high fever as well as the development of petechiae (hemorrhagic areas under the skin). Only a small number of cases (fewer than one percent) go on to develop the more serious viral hemorrhagic fever syndrome, leading to death in fifty percent of those who manifest this syndrome. This syndrome is associated with mucosal bleeding or hemorrhaging, liver and kidney failure, and shock before death. Some infections can be complicated with encephalitis and a variety of ocular (visual) defects.

Diagnosis

While the VHF syndrome experienced by RVF infected patients is not specific for RVF, there are some manifestations that are: the occurrence of an epidemic febrile hemorrhagic disease with eye lesions and encephalitis experienced by some members of the infected population would, from an epidemiological (study of epidemics) view, be most characteristic of RVF.

- *Laboratory* findings would be similar to those for other VHFs, indicating a reduction in the number of white blood cells as well as abnormal liver chemistries (elevated).

- *Special tests:* The presence of viruses can be detected in blood by ELISA testing. In addition, antibodies to RVF can be detected by similar ELISA methods.

Treatment

As with all VHFs, supportive therapy must be given dependent upon the complications experienced by patients. Patients would be treated with ribavirin intravenously for four to six days. While there are no human studies at this time to verify the efficacy of this treatment, there are both cell studies and rodent animal studies that attest to the efficacy of ribavirin. There is an effective inactivated vaccine available that can be administered in three doses. Protective antibodies appear before fourteen days and last one year. Hence annual boosters must be given.

PROTECTIVE RESPONSE STRATEGY

- The strategies for CCHF would also be applicable to RVF.

- Any individual who may have been exposed to blood, body fluids, secretions, or excretions from a patient with a suspected VHF should immediately and thoroughly wash the skin surfaces with soap and water.

- In hospitals, contact precautions should be instituted to prevent contagion. A special respiratory mask called N95 is recommended and should be worn for protection. Infected patients should be quarantined for the duration of the illness.

- Anyone caring for (or visiting during convalescence) a patient who is stricken with a VHF must practice strict "barrier nursing" techniques (in other words, completely protect oneself from the infected patient by dressing in gown, gloves, mask, hat, booties, etc.), because evidence has accumulated indicating that large droplets or even fomites (inanimate objects) may act as mediators of transmission of the virus.

- Any contamination of mucous membranes must be immediately diluted with copious quantities of water. Contamination of eyes can be diluted with saline or OTC products like Visine.

- Decontamination of surfaces, equipment, or other articles can be accomplished by using a ten percent solution of Clorox. Lysol-type disinfectants are an alternative.

- Deceased individuals should be sealed in leakproof material for prompt burial or cremation.

- VHFs (including RVF) are very difficult for a bioterrorist to weaponize, and there is no real carrier state for VHF, making it an unlikely weapon.

• Articles of clothing should either be discarded if heavily soiled, or washed well using a strong detergent containing a germicide, or washed using bleach.

15.

RICIN

Bioterror Agent	Ricin
Type of Weapon	Toxin of *Ricinus communis,* castor bean plant
Disease Caused	Poisoning
Transmissible (person-to-person)	No
Incubation Period	2 to 8 hours
Cardinal Features of Disease	Abrupt-onset nausea and vomiting, abdominal cramps, diarrhea, chest tightness, vascular collapse, shortness of breath
Treatment	Supportive
Vaccines Available	Yes (experimental)

HISTORY AND BACKGROUND

Ricin is a very potent protein toxin derived from the mash left over from processing castor beans for their oil. Such processing is a worldwide activity and the toxin is easily produced. The castor plant, *Ricinus communis,* received attention in the 1978 "umbrella murder" case, when Bulgarian intelligence operations assassinated a Bulgarian dissident, Georgi Markov, in London using an umbrella to deliver a ricin-tipped bullet. Mr. Markov died a day after the attack of ricin poisoning. Ricin blocks protein synthesis and thus is very toxic to cells, eventually killing them.

If a bioterrorist were to use ricin as a weapon, it would probably be aerosolized, but it could, alternatively, be used by an assassin, who might prepare the toxin for ingestion or injection. The toxin is made up of two polypeptide chains, an A chain and a B chain, which are bonded together. Even though it is a potent toxin, it is not nearly as potent as the botulinum toxin, and thus it would have to be produced in very large quantities for large-scale use. Ricin is a convenient toxin, because it can be prepared as a liquid, as a crystal, or even lyophilized (dried) as a powder. Within eight hours of *inhalation* as an aerosolized small particle, ricin will cause severe respiratory symptoms leading to a respiratory failure in two to three days. If the ricin toxin is *ingested* (food or water), severe gastrointestinal symptoms will occur, followed by vascular collapse and death. If ricin is *injected* into a victim, besides vascular collapse, it can cause multiple organ failure, resulting in death. This information was gleaned from studies done in animals, particularly rodents. In rodents ricin was shown to be more toxic by aerosolization than by any other route of exposure. While there is very little data on ricin use in humans, it is presumed that an aerosol exposure in man would cause significant lung damage and pulmonary edema (fluid buildup in the lungs). This would eventually lead to respiratory failure and cardiac arrest. In rodents that were exposed to ricin via an aerosol, their respiratory tract, including their trachea, bronchi, and lungs, showed necrotizing lesions with edema. Ricin poisoning is not transmitted person-to-person.

CLINICAL FEATURES

Obviously the clinical picture of ricin poisoning would depend upon the route of its administration. All serious cases share at least the following symptoms: abrupt onset of nausea and vomiting, abdominal cramps, diarrhea, chest tightness, vascular collapse, shortness of breath (dyspnea), and arthralgias. Toward the termination of these symptoms, a profuse sweating episode completes the picture. Such a clinical picture was first recorded after an accidental sublethal aerosol was inadvertently launched on humans. If ricin were delivered via ingestion or intramuscular

(IM) administration, there would not be a severe lung involvement. As previously discussed, ingestion causes GI hemorrhage with hepatic, spleenic, and renal necrosis. If ricin is injected IM, severe local necrosis of muscle and lymph nodes occurs, with some organ damage.

Diagnosis

Differential diagnosis would require ruling out SEB, Q fever, plague, and tularemia, and several chemical agents such as phosgene. Each of these agents can cause severe lung injury. One key observation that would be made during a bioterrorist attack with ricin would be the large cluster of cases involving individuals with the same type of pulmonary symptoms. Chemical nerve agents like phosgene gas would act much faster to produce an immediate type of lung injury that would be differentiated from the delayed (eight-hour) type caused by ricin.

- *Laboratory:* It appears that the white blood cell count rises after about twelve hours following aerosol exposure to ricin. This is based on laboratory animal studies, but it is assumed that this would also occur in humans. As part of the differential diagnosis for ingested ricin, infection caused by enteric (gastrointestinal) pathogens would have to be made. While both enterics and ingestion ricin poisoning would have a gastrointestinal involvement, only ricin poisoning would feature vascular collapse as a prominent symptom. That would likely cast suspicion away from enteric pathogens like *Salmonella, Shigella,* and *Campylobacter.*

- *Special tests:* Blood can be used to detect the presence of ricin protein toxins by ELISA methodology. In addition, punch biopsies of tissue lesions can be analyzed using immunohistochemical techniques to confirm ricin poisoning. Lastly, because ricin is a protein it is immunogenic (induces antibody production) and thus causes the development of antibodies, which can be detected quite readily. PCR is also available to detect contaminating castor bean DNA.

Treatment

Supportive management of patients depends upon the route of administration of ricin and therefore is varied. Currently there is no vaccine available for prophylaxis, nor is there an antitoxin available for treatment. However, both are under development. Preliminary studies show efficacy in animal studies with these experimental vaccines.

PROTECTIVE RESPONSE STRATEGY

- Ricin toxin is not dermally active and can be removed from the skin with soap and water.

- A mild bleach solution (ten percent bleach) can effectively inactivate ricin on surfaces and fomites.

- A protective respiratory mask, if worn properly, is effective at preventing ricin poisoning.

- Ordinary clothing would provide a body shield from toxin, and routine washing of clothing would effectively remove toxin. There is no need for special protective clothing. All non-clothing articles such as watches, rings, eye-glasses, jewelry, wigs, hearing aids, and other items not part of the body should be soaked in a ten percent bleach solution for about ten to thirty minutes, then flushed with water and dried before reusing.

- Even if a bioterrorist contaminated the water supply with ricin, municipal water supplies are very difficult to contaminate sufficiently to cause significant casualties for three reasons:

  1. Dilution and diffusion factors.

  2. Chlorination inactivates toxin.

3. Reverse osmosis systems utilized in municipal water treatment plants are effective against most toxins.

- Toxins are not very effective aerosol weapons. Open-air exposure limits effectiveness because of the dilution factor.

- For ingested toxins, activated charcoal tablets might be administered to the victim if he or she is conscious and alert.

16.

SAXITOXINS

Bioterror Agent	Saxitoxin
Type of Weapon	Toxin from dinoflagellate protozoa
Disease Caused	Poisoning
Transmissible (person-to-person)	No
Incubation Period	Minutes to an hour
Cardinal Features of Disease	Numbness of lips, tongue, and fingertips; muscular incoordination, "floating" sensation
Treatment	Supportive
Vaccines Available	No (experimental anti-toxin)

HISTORY AND BACKGROUND

Saxitoxins include many chemically related neurotoxins primarily produced by marine plantlike proto-
zoa called dinoflagellates. Humans acquire such
toxins by eating bivalve mollusks that have been
feeding on dinoflagellates. This results in a specific
syndrome called paralytic shellfish poisoning (PSP),
which is a life-threatening medical condition re-
quiring immediate treatment. In nature PSP is a relatively rare syndrome. It is most often associated
with "red tide" (a summer algal bloom). The disease
is most often seen in the U.S. along the Saint
Lawrence seaway region and in New England and
Pacific coastal states. Saxitoxins are not destroyed by
heat, and symptoms of paralytic poisoning occur
within thirty minutes of consumption of contaminated shellfish. Within minutes (paralytic type) or
hours (neurologic type) a victim can experience burning of the mouth and extremities, headache, vertigo,
nausea, vomiting, diarrhea, muscle weakness, or
paralysis. Symptoms are milder in the neurotoxic
type. The saxitoxins are water-soluble compounds
that block nerve-to-muscle transmission. In other
words, they prevent proper nerve functioning.
Death usually results from respiratory paralysis. A
bioterrorist would probably deliver his weapon aerosolly or perhaps use it as a poison to contaminate
food or water.

CLINICAL FEATURES

Onset of symptoms occurs within minutes to an hour after ingestion or inhalation, dependent upon the quantity of toxin taken in. The first set of symptoms include numbness or tingling of the lips, tongue, and fingertips. Numbness progresses to the extremities with or without a burning sensation. This is followed by muscular incoordination. Nausea and vomiting occur only in a minority of cases. Some individuals report a "floating" sensation, generalized weakness, dizziness, inability to speak properly, memory loss, and headache. Patients remain conscious. Those who survive the first twelve to twenty-four hours usually recover, because the toxin is cleared quickly by the body.

Diagnosis

The differential diagnosis involves ruling out other toxin-mediated diseases, such as ciguatoxins (ciguatern toxins), which cause the most common fish-borne illness worldwide, and the most common nonbacterial food poisoning reported in the U.S. The ciguatoxins are also created by single-celled organisms living on coral reefs. These organisms attach themselves to coral reef algae, which are eaten by small fish, which in turn are eaten by predator fish; thus the toxin starts to accumulate in these latter fish, especially those caught off the Caribbean coral reefs, such as grouper, snapper, sea bass, and barracuda. People who eat these larger fish get a dose of toxin. The differential characteristic between saxitoxins and ciguatoxins is that there is a much greater degree of GI involvement with ciguatoxins. Also, some pesticides can cause a poisoning similar to these toxins, but that can be identified and differentiated by using a gas-chromatography analysis of the stomach contents (or the food in question).

- Laboratory: Routine laboratory tests are not really helpful in identifying a toxin-mediated illness. However, there are numerous special toxin assays that can be used to detect saxitoxin as well as others by the ELISA technique.

Treatment

As with other toxin therapy, treatment is limited mainly to supportive management of patients. Mechanical respiration support may be required in severe cases. There is an antitoxin that has shown success in animal models, but to date there is no human data. No vaccine against saxitoxin has been developed as of yet. Induction of vomiting may prove to be useful as part of the treatment program. Ipecac syrup is a good emetic that could be used to induce vomiting.

PROTECTIVE RESPONSE STRATEGY

- Saxitoxins are water-soluble and can be removed from the skin with soap and water.

- A mild bleach solution (ten percent bleach) can effectively inactivate the saxitoxin on surfaces and fomites.

- A protective respiratory mask, if worn properly, is effective at preventing saxitoxin-inhalation poisoning.

- Even if a bioterrorist contaminated the water supply with saxitoxin, municipal water supplies are very difficult to contaminate sufficiently to cause significant casualties for three reasons:

  1. Dilution and diffusion factors.

  2. Chlorination inactivates toxins.

  3. Reverse osmosis systems utilized in municipal water treatment plants are effective against most toxins.

- Clothing would provide a protective body shield from toxin and routine washing of clothing would remove toxin. All nonclothing articles such as watches, rings, eyeglasses, jewelry, wigs, hearing aids, and other items not part of the body should be soaked in a ten percent bleach solution for about ten to thirty minutes, then flushed with water and dried before reusing.

• Toxins aren't very effective aerosol weapons. Open-air exposure limits effectiveness because of the dilution factor.

• Activated charcoal tablets might be administered to the victim if he or she is conscious and alert.

17.

TRICHOTHECENE MYCOTOXINS (T-2)

Bioterror Agent	T-2 (Trichothecenes)
Type of Weapon	Fungal toxins
Disease Caused	Poisoning
Transmissible (person-to-person)	No
Incubation Period	2 to 4 hours
Cardinal Features of Disease	*Ingestion:* nausea, vomiting, diarrhea *Pulmonary:* difficulty breathing, coughing, wheezing *Systemic:* generalized weakness, dizziness, low blood pressure, rapid heart-beat
Treatment	Supportive
Vaccines Available	No (experimental antitoxin)

HISTORY AND BACKGROUND

The trichothecene mycotoxins (T-2) are very low-molecular-weight compounds produced by certain fil-amentous fungal molds of the following genera:

Fusarium

Myrotecium

Trichoderma

Stachybotrys

There are about 150 trichothecene compounds that have been described in the literature. They are insoluble in water but soluble in alcohol. While they are very stable to heat ($1500°F$ for thirty minutes is needed to inactivate) and UV light inactivation, they are easily destroyed in the presence of a mild bleach solution.

Mycotoxins gained notoriety during biological war-fare incidents in Laos and Afghanistan (1979–81) in which an estimated 6,300 and 3,042 deaths occurred, respectively. The agents of these biowarfare events were nicknamed "yellow rain."

The trichothecene mycotoxins inhibit protein syn-thesis, impair DNA synthesis, and interfere with cell membrane structure and functions. In nature, T-2 tox-ins affect humans and animals by inhalation or con-sumption of contaminated food products. For example, a natural disease in cattle can occur if animals are fed corn that has been contaminated with a toxin-producing fungus. There are three avenues of toxin poisoning: aerosols, ingestion, and skin absorption. A bioterrorist might employ a weapon that might be delivered by any of these avenues.

CLINICAL FEATURES

T-2 toxins are able to enter the body through the skin, digestive, or respiratory tissues. The toxins act very quickly on tissues of the skin, mucous membranes, and bone marrow. These toxins are different from most in that they can adhere to and penetrate the skin. They can also be very effective if inhaled or ingested. Hence, T-2 toxins should be treated more like chemical rather than biological agents. Simple precautions (such as using ordinary face masks) used for the latter are insufficient protection against T-2 toxins.

Ingestion of T-2 mycotoxins can result in nausea, vomiting, and diarrhea (sometimes bloody), with abdominal cramping. Mouth and throat pain is evident, and saliva or sputum can be blood-tinged. Eye pain, tearing, and blurred vision can also occur.

Pulmonary (inhaled) T-2 is usually characterized by difficulty breathing with wheezing and coughing. Nasal intake results in pain and sneezing with nasal bleeding.

Systemic toxicity can result if sufficient T-2 mycotoxins are either ingested or inhaled. Such toxicity manifests as generalized weakness, dizziness, overall loss of coordination, low blood pressure, rapid heartbeat (tachycardia), and low body temperature (hypothermia). Death can occur in minutes or days.

Diagnosis

The differential diagnosis involves ruling out other toxin diseases as well as radiation sickness.

- *Routine laboratory* testing shows an increase in white blood cells (leukocytosis). Other lab tests are not significant.

- *Special tests:* Must be left to specialized biochemistry labs that possess gas-liquid chromatography machines to detect T-2 or related toxins in blood and urine of victims. Eventually fifty to seventy-five percent of toxin is eliminated in urine and feces of victims within one day. Urine is the ideal specimen to process.

Treatment

"Superchar," an oral superactivated charcoal preparation, is standard therapy for poison ingestion. This type of charcoal preparation is five times stronger than ordinary charcoal.

Certainly supportive therapy must be provided to victims dependent upon the symptoms displayed.

There is no antitoxin currently available for human use. However, there is an experimental antitoxin that has shown some efficacy both therapeutically and prophylactically in animal models.

PROTECTIVE RESPONSE STRATEGY

- Although T-2 toxins are insoluble in water, simple soap and water can easily remove them from the skin. For skin inactivation an alternative to removal with soap and water is to use a mild bleach solution (ten percent) mixed with three percent hydrogen peroxide.

- A mild bleach solution (ten percent) is necessary to effectively inactivate T-2 toxin on surfaces and fomites.

- Even if a bioterrorist contaminates the water supply with T-2 toxins, municipal water supplies are very difficult to contaminate sufficiently to cause significant casualties for three reasons:

  1. Dilution and diffusion factors.

  2. Chlorination inactivates toxins partially or completely.

  3. Reverse osmosis systems utilized in municipal water treatment plants are effective against T-2 toxins.

- A protective respiratory mask (with activated charcoal), if worn properly, is effective at preventing T-2 toxin inhalation poisoning.

- Any contaminated clothing might best be discarded. If possible, all nonclothing articles should be soaked in a ten percent bleach solution for thirty minutes, then flushed with water before reusing.

- Toxins aren't very effective aerosol weapons. Open-air exposure limits effectiveness because of the dilution factor.

18.

CHEMICAL AGENTS AND POISON GASES

Bioterror Agent	Chemical agents and gases like sarin
Type of Weapon (many other varieties)	Chemical nerve agent
Disease Caused	Poisoning
Transmissible (person-to-person)	No
Incubation Period	Immediate
Cardinal Features of Disease	Varied, muscular twitching, pinpoint pupils, rhinorrhea, salivation, tightness of breath, neurologic symptoms (paralysis, loss of consciousness)
Treatment	Supportive and antidotes
Vaccines Available	No

HISTORY AND BACKGROUND

In the previous sections, I discussed three categories of biological weapons: deadly bacteria, viruses, and toxins produced by germs. In this section, I'm presenting chemical agents and poison gases that might be used by a terrorist. These agents have been placed into thirteen different categories by the CDC. The following is a list of the categories, with one example for each category:

1. Nerve agents (sarin)

2. Blood agents (hydrogen cyanide)

3. Blister agents (nitrogen mustard)

4. Heavy metals (arsenic)

5. Volatile toxins (benzene)

6. Pulmonary agents (phosgene)

7. Incapacitating agents (benzilate)

8. Pesticides (malathion)

9. Dioxins

10. Explosive nitro compounds (ammonium nitrate)

11. Flammable industrial chemicals (gasoline)

12. Poison industrial chemicals (cyanides)

13. Corrosive acids and bases (sulfuric acid)

There are literally hundreds of chemical agents and poisonous gases that can be used in a bioterror attack. Hundreds more new chemicals are introduced every year that are unknown to us yet we may face in a future bioterror attack. Many of these agents may be derived from newly described genetic engineering methods, which actually will allow the creation of an infinite number of possibilities. As such, medical and first-responder personnel must react to emergencies and base their treatment of victims of any attack on syndromic categories, such as burns, cardiorespiratory failure, neurologic damage, etc.

It is not within the scope of this monograph to describe most or even some of these chemical agents. One such chemical—sarin—will be presented as an example for discussion.

Sarin is a toxic organophosphorus nerve agent, which is odorless, colorless, and tasteless and diffuses very rapidly into the human skin because of its high volatility (ease of evaporation). Sarin is twenty-six times more deadly than cyanide gas, and twenty-one times more lethal than potassium cyanide. Once released, sarin tends to drift above ground for weeks or even months. The mechanism of action of nerve agents such as sarin is that they disrupt nerve communication with the organs they stimulate. In other words, the nerve is normal but the transmission of the nerve impulse to the muscle or other organ is faulty, causing overactivity. Sarin is extremely toxic by any routes of exposure. If heated or if it starts decomposing it can emit very poisonous fluoride and phosphorus oxide fumes. Death can occur as soon as one to ten minutes after inhalation. Nerve agents share some other interesting properties. If they are in liquid form they are heavier than water. Their vapors are also heavier than air, and therefore they would sink toward the ground or the basement of a building. People actually use weaker forms of nerve agents every time they tend a garden with insecticides. For example, insecticides like malathion or sevin and dozens of others are nerve agents. They do the exact same thing to humans that sarin or other nerve agents do, except that it would take much larger doses and longer contact time to have a similar effect.

Although nerve agent vapor will have effects on victims in a very short period of time, the range of these effects will vary greatly depending upon the degree of that exposure. In the 1995 Tokyo subway system attack (outlined in the Introduction), the injuries ran the gamut from very minor (effects experienced by seventy-five percent of the casualties) to fatalities. Numerous people who inhaled sarin gas on the train collapsed. When the final tallies were in, twelve people died and over five thousand were injured to different degrees. Just about every person exposed to sarin shared one particular symptom— miosis (small pupils). Miosis was usually bilateral; rarely was it unilateral.

CLINICAL FEATURES

Exposure to nerve agents will initially affect airways and the portions of the face that come into contact with the agent: the eyes, nose, and mouth. The pupils become small, the eyes become reddened, and vision becomes blurred. Some patients also experience eye pain, headache, and nausea and vomiting. Rhinorrhea (runny nose) may be an important feature, as is excessive salivation. If sarin in inhaled, airways can become constricted and induce coughing fits or shortness of breath. If sufficient sarin is inhaled, there can be a sudden loss of consciousness followed by convulsions. Within a few minutes a victim might stop breathing and then become flaccid (completely limp). Even a very small amount of sarin on the skin can produce a phenomenon called fasciculations (muscular twitching) at the drop site. Nausea and vomiting may accompany any exposure to sarin. In cases of severe exposure, involuntary defecation and urination may occur.

Diagnosis

The first indication of exposure to sarin may be a specific reaction at the point of contact of the nerve agent. For example, symptoms like muscular twitching, pinpoint eye pupils, rhinorrhea, or tightness of breath coupled with shortness of breath might indicate poisoning by a nerve agent. The other telltale sign would be the rapidity of the reaction—in minutes after exposure.

- *Laboratory:* First-responder teams are
  equipped with a variety of chemical moni-
  toring kits that can indicate the presence or
  absence of a suspected class of chemical at
  the site of an event. It is important to make
  an immediate identification of a chemical
  agent, because time is of the essence when
  dealing with such agents. Any delay can lead
  to unnecessary death if the appropriate anti-
  dote is not given. The type of chemical
  detection equipment used by hazmat (hazardous materials) teams may vary depending
  upon the locality or the event. Many
  machines are portable and can be operated
  at the site; others are too sophisticated for
  use anywhere but in a laboratory. The types
  of instruments range from spectrometry and
  chromatography to photoionization detec-
  tors and simple color-change chemical
  methodologies. There are also a wide assort-
  ment of biochemical blood tests available at
  medical facilities that can be used as confir-
  matory tests or when patients are brought
  into hospitals for treatment.

Therapy

The immediate treatment involves giving antidotes to
victims as soon as the chemical agent is identified. For
example, antidotes for nerve agent poisoning are

atropine (which blocks the effects of the chemical that causes overstimulation of nerves) and 2-PAM chloride (which removes the nerve agent from the enzyme that the nerve agent interferes with). Of course, ventilation and oxygen should be provided if necessary. An anticonvulsant called diazepam may be given if a patient is convulsing. Variation to these medications would obviously occur depending upon individual circumstance. Simply put, the ABCs of victims must be managed (airways, breathing, and circulation).

PROTECTIVE RESPONSE STRATEGY

- Responders to a chemical-agent attack must be protected with special masks, gloves, and protective suits until victim is decontaminated.

- Remove victim from contaminated area, and then remove contaminated clothing from victim. Clothing should be bagged and later destroyed.

- Decontamination of any victim can be done by showering with soap and water and dilute (ten percent) bleach solution (applied directly to the victim's skin).

- Eyes must be flushed with warm water for fifteen minutes if exposed to sarin.

- Victims should be removed to fresh air as soon as possible.

- Emergency personnel should avoid self-contamination.

- Never perform mouth-to-mouth resuscitation especially when facial contamination is known to exist.

- Never induce vomiting.

- Activated charcoal might be administered to victim if he or she is conscious and alert.

- Hair should be shaved if contaminated by sarin. It is to be bagged and later discarded.

- Do not rush to aid victims unless you are fully protected with a suit.

EPILOGUE

FACING THE FUTURE, POST-9/11

More than any other event, the World Trade Center disaster and the subsequent anthrax attack on the United States dramatically and clearly demonstrated the grave danger that acts of terrorism are to the civilian population. In reality there are an infinite number of diabolic ways that a motivated terrorist could send his message or make his statement. Well before the events that unfolded in the aftermath of the World Trade Center disaster, I wrote a vignette in my book *The Secret Life of Germs* about a disgruntled employee who decided to make such a statement. Little did I realize how prophetic that vignette would turn out to be:

It hasn't happened—yet. But consider the following scenario.

December 31, 11 P.M.: In a deserted hangar at the Old Rhinebeck Aerodome in Rhinebeck, New York, about a hundred miles up the Hudson River Valley from New York City, a man readies a crop-dusting plane for a special New Year's Eve flight. Midway through his preflight checklist he takes a vial from his pocket and swallows his third dose that evening of doxycycline, an antibiotic, as a prophylactic measure. He doesn't really fear being infected tonight; his big risk of exposure is past. And ultimately he doesn't care when he goes down in flames. But he is determined to live long enough to enjoy the statement he's about to make.

It's a statement he's been planning for two years now, ever since his dismissal as a civilian laboratory technician from the U.S. Army's Medical Research Institute of Infectious Diseases at Fort Detrick, Maryland. His supervisor said he had been "consistently lax in the handling and treatment of hazardous materials." That was a lie. Well, now he'd show that jerk supervisor and everybody else in the lab that he knew what he was doing. Now he'd show the whole world.

Deciding what to do and how to do it has been easy. The Army's own antibioterrorism manuals had given him the basic concept, although when and where were strictly his idea. Fortunately, he had a little capital to work with. His parents' dinky house in Camden, New Jersey, had been sold the year before, after his mother died, and the money was still sitting in the bank. It was enough to live on, if he was frugal, until he put his plan into effect. After that, money wouldn't matter anymore.

The tough part had been getting his pilot's license and, even more of a problem, getting access to a crop-dusting plane. He earned his license a year after being canned at Fort Detrick. Then he spent months trying to hire on with one of the handful of outfits that did crop dusting in the New York, New Jersey, Connecticut area. For a while it seemed as if he might have to wait another whole year to execute his plan. Then a pilot for Catskills Crop Dusting took off for parts unknown just as the busy season was heating up, and a job opened up for him. It was like a sign from fate that his plan couldn't fail.

One last check of the instrument panel and he's ready for takeoff. He wishes he could fly straight over New York City and the party in Times Square, but that's a little too risky. Besides, the real fireworks will come in a few days, if the germs get a chance to do their stuff.

He feels like it's up to him to see that they do.

He flies south to within twenty miles of the city and then circles around to the northeast, where the prevailing winds are coming from that night, exactly as forecast. He turns the plane into the wind, checks his course, and then, with the lights of the city glittering behind him, releases the crop duster's payload. He giggles as two hundred pounds of powdered anthrax begin floating down toward the city. The particles are very small, five microns or less in diameter, so that gravity doesn't drag them down too quickly. By the same token, the nighttime temperature inversion will keep the particles from rising too high in the air. By the time they reach New York City, they should be settling down to just about street level.

It had been a snap to get the anthrax. Four months before his dismissal, a batch of samples had come in from the Middle East, where the soil commonly contains anthrax bacilli. He'd been assigned to work with the samples, isolating the microbes and growing out colonies to check that they were susceptible to ciprofloxacin and doxycycline and to see if any mutant strains had occurred in the wild. He slipped a couple of tablespoons' worth of soil into his pants pocket and took it home to culture anthrax bacilli in petri dishes on his kitchen windowsill. At first he hadn't known what he was going to do with the anthrax. But it was exciting to have, better than any gun. He liked to imagine what it could do if it were released in the air-conditioning vents of a shopping mall or dropped over a St. Patrick's Day parade, or best of all, scattered over Times Square in New York City on New Year's Eve. The whole world would see a mass murder take place, and only realize it after the fact!

Then they fired him, and that became his mission. With his lab skills it was simple enough to grow the anthrax colonies he needed, subject them to desiccation so that they would sporulate, and then grind the spores up into powder. While doing this work, he made sure to wear protective gear and dose himself with doxycycline or ciprofloxacin, also courtesy of the lab at Fort Detrick. He'd been helping unpack a shipment of supplies when he got the chance to steal the antibiotics, stashing them behind the cabinet and then announcing that the shipment and the packing slip didn't agree. Suspicions about that had probably helped get him fired. But one of the best parts of the whole scheme was all the flak his old bosses were going to catch for being "consistently lax in the handling and treatment of hazardous materials." He was going to get a big kick out of that.

Ciprofloxacin and doxycycline are the primary

antibiotics that offer protection against infection with anthrax bacilli, if they are taken in advance of, or very soon after, exposure. The man in the plane doubts that any more than a handful of medical experts and counterterrorism personnel will be on the right antibiotics on this New Year's Eve. For the thousands of people celebrating in Times Square or out on the streets elsewhere as the anthrax bacilli float down, inhaling them will be an almost certain death sentence. After a one-to-six-day incubation period, infected people will begin to feel as if they are coming down with the flu. If any of them are lucky enough to be treated with ciprofloxacin or doxycycline—a remote possibility, he figures—they might pull through. But the odds are that suspicions of anthrax won't be aroused until the next day. Then people will begin to suffer respiratory distress, in many cases with a telltale pain in the middle of the chest called mediastinal involvement. As clusters of people with these symptoms turn up in emergency rooms all over the tristate area, some alert doctor, nurse, or public-health officer may raise an alarm about a possible outbreak of anthrax infection. But by then toxins will be coursing through the victims' bloodstreams, and it will be too late to save their lives. Within about twenty-four hours they will go into shock, and shortly afterward they will die.

The man in the plane can see it all happening in his mind's eye. He can't wait to see the video on CNN. Sooner or later, he knows, the authorities will get on his trail. But that doesn't matter. They could arrest him in midair, for all he cares.

Okay, my plot is full of stretches and coincidences, and I'm obviously no threat to John le Carré, but I've spun the scenario out at this length in order to highlight some of the most important issues in bioterrorism and biological warfare. Anthrax could indeed be effectively delivered with a crop duster, and if it were—by a disturbed individual or a team of terrorists—it could exact a very heavy toll. Congress funded a study that reported that two hundred pounds of powdered anthrax spores floating down on the Washington, D.C., area would kill up to three million people. A cigar box can hold enough anthrax to kill tens of thousands of people. I believe that in order to achieve that death rate a government would have to sponsor and execute the terror attack.

More realistically, the Russians have had some unfortunate real-life experiences that show how easily people can be killed by the release of anthrax bacilli into the air. In 1979, workers at an anthrax plant in Sverdlovsk, in central Russia, unwittingly sprayed anthrax particles into the outside air when they operated a drying machine with a missing filter. About four hundred people, some living three miles away, were made sick, and about ninety people died. A flock of sheep grazing some thirty miles away also died. In 1992, Boris Yeltsin admitted that sixty-six people living downwind from a Russian government lab had been killed by another accidental release of anthrax.

In spite of the "real" anthrax attack against the U.S. that occurred in September 2001, my "imaginary" anthrax scenario is possible but not very probable. In order for a crop duster to effectively deliver the proper quantity of particle-sized spores, the sprayer heads would have to be retrofitted. Such retrofitting could, in fact, be performed, but only after considerable effort. I believe that, at this time, the state of vigilance of the American populace is unprecedented and would in and of itself be a deterrent. Having about three hundred million pairs of "watchful" eyes is an awesome **Protective Response Strategy,** which can limit a bioterrorist's activities. However, my vignette did exemplify the fact that bioterror is the domain not only of a government or a rogue state but also of an organization, or even a highly motivated disgruntled individual without any connection to anyone or anything. While no one can prevent a suicide bomber determined to kill himself along with others in his proximity from executing that plan, with a concerted effort we as a people, along with our government, can limit the effectiveness of any terrorist by a sustained and concerted cooperative effort. Despite the fact that, in past years, industrialized nations including our own have been discussing bioterrorism preparedness, those discussions have always been halfhearted, mainly because of strong feelings of disbelief that any of us were vulnerable. Nevertheless, we have indeed experienced large-scale "bioterrorism-like" outbreaks of disease, which have inadvertently allowed us the opportunity to practice for a major attack.

Well before September 2001, a municipal water supply became contaminated and an entire city was affected. That happened in Milwaukee, Wisconsin, in 1993 because of *Cryptosporidium,* a parasitic protozoan germ commonly found in surface water. It usually gets there through feces of birds, including geese and ducks, reptiles, and many other animals, including human beings. Some four hundred thousand people, over half Milwaukee's population, became ill with profuse watery diarrhea and abdominal pain; one hundred people died. Since then numerous additional episodes of cryptosporidiosis in the United States have been traced to drinking supposedly safe, treated water from the tap. The scope of the problem has become so great that in 1998 the federal government announced an $800 million water-treatment plan to safeguard the nation's drinking water. This outbreak could have just as easily been caused by a bioterrorist!

Health authorities were faced with an unprecedented major health crisis; they investigated it epidemiologically, they identified its cause, and they provided a solution within a reasonable time frame. They met the

challenge in much the same way our public-health system in conjunction with law-enforcement agencies
would respond to a bioterrorist event. One year after
the waterborne cryptosporidial outbreak, a national
food-borne outbreak of *Salmonella* occurred when
250,000 people consumed contaminated ice cream.
Both of these major events were learning experiences
for both public-health and law-enforcement agencies.

It may be hard to believe that a health system that
couldn't get its influenza vaccine to the general public
in time for the flu season in the year 2000 faced an
unprecedented anthrax bioterrorism attack in 2001
and met that challenge. But in fact, the response by all
involved was exemplary and allowed both health and
law-enforcement networks an opportunity to get their
acts together and prepare for just about any major
future terrorist attack. In a sense the limited anthrax
attack was like a practice run for a larger event.

The eighteen likeliest bioterror agents presented in
this compendium each pose specific potential risks to
the general population. But let's examine closely what
those risks really are.

*First, poisoning water supplies with any agent will
never be a major threat.* Any contamination of the
water supply is difficult for three reasons:

- Large bodies of water can dilute a contami-
  nant sufficiently well to make it relatively
  ineffective.

- Chlorination inactivates most toxins and kills many harmful bacteria and viruses.

- The process of reverse osmosis utilized in most municipal water treatment plants effectively removes unwanted chemicals and poisons.

Some systems even use ozonation (superoxygen), which kills the heartiest of germs. The same limitations apply to poisoning foodstuffs. There would be limited effectiveness. Most of the biologic agents that would be added to food would not cause significant mortality. It can be said that no matter the method of delivery of a bioterror agent, only small-scale attacks have shown to be successful in the past. Such was the case for the letters laced with anthrax spores delivered to the media and senators Daschle and Leahy. Even the remains of charred papers with instructions for chemical and biological weapons, found in a compound in Afghanistan used by Osama bin Laden's network, Al Qaeda, don't change the picture: Developing biological weapons without having an appropriate delivery system is like having bullets without a gun.

What about a smallpox attack? What would be the likelihood of it spreading like wildfire throughout America? The short answer is: not very likely! Smallpox is far from a perfect weapon. It is assumed that only the U.S. and Russia have the smallpox virus.

but it is entirely possible that a rogue state has gotten the virus through illicit means. Let's say a few cases of smallpox do surface. According to the CDC only one or possibly two secondary cases would occur for each primary case. Previously it was thought that one case could infect hundreds. In addition, there is a large window of opportunity of four days for appropriate treatment with a vaccine in order to prevent infection after exposure. There are about 15 million vaccine doses currently available, which could be diluted to accommodate a total of up to 150 million people (half of our entire U.S. population). The U.S. government has just contracted for the purchase of about 155 million new doses of smallpox vaccine. Even those patients who lose the window of opportunity can be treated with VIG (vaccinia immune globulin), which will neutralize the smallpox virus until their vaccine kicks in. It is also entirely possible that most of those individuals who were vaccinated against smallpox still have circulating within their bodies "memory immune cells" that will increase upon re-exposure to the virus and thereby prevent full manifestation of disease. There are even antiviral drugs such as cidovir that show promise. It is also entirely possible that most of those individuals who were vaccinated against small-pox still have circulating within their bodies "memory" immune cells that will increase upon re-exposure to the virus and thereby prevent full manifestation of disease. In my opinion, we can prevent a widespread smallpox outbreak in a way similar to the methods used to eradicate smallpox from the face of the Earth!

Certainly history records that tens of millions of people died of the plague or Black Death *(Yersinia pestis)* from the Middle Ages to the mid-nineteenth century because untreated "pneumonic" plague had a mortality rate of one hundred percent and "bubonic" plague has one of about fifty percent. However, with antibiotic treatment (something that wasn't possessed by early civilizations) it has a very low death rate. In addition, we now have appropriate means to control rodents and fleas, which would also be needed during a plague outbreak.

What about brucellosis ("undulant fever") or tularemia ("rabbit fever")? The fact is that neither is a good bioterror weapon, because they both have a low fatality rate and with adequate antibiotic therapy that fatality rate is reduced to almost zero. A similar situation exists for both Q fever *(Coxiella)* and cholera *(Vibrio)*, because these diseases are also rarely fatal. If used at all, they would probably be used against soldiers in the battlefield in order to incapacitate them or compromise their fighting performance. Melioidosis and glanders would likely be weaponized for an aerosol delivery. However, as with many agents, both are not good candidates for bioterrorism because they are detectable by ordinary bacteriological media (hence they are easily identifiable) and are treatable with antibiotics, which would reduce the relatively high death rate to near zero. These organisms would also probably make better military weapons. On the battlefield a delayed detection might lead to a higher death rate.

Although there are many types of encephalitis viruses available for a weapon, Venezuelan equine encephalitis (VEE) is the top choice even though it has a fatality rate of less than one percent. The reason is because one hundred percent of humans exposed to VEE will develop an influenza-like illness. Again such a weapon would also be better suited as an incapacitating agent on the battlefield.

Viral hemorrhagic fevers such as Crimean Congo and Rift Valley fever and Ebola are pretty horrific diseases. Yet they are difficult for a bioterrorist to weaponize, and there is no real carrier state for VHF, making it an unlikely weapon. The hemorrhagic diseases can also be contained by the practice of strict "barrier nursing" techniques.

Ricin is a convenient toxin because it can be prepared as a liquid, a crystal, or a powder. It is likely that this toxin would be delivered in small-scale attacks. It is more a boutique-type weapon. The same is true of saxitoxin, T-2 toxins, and other toxins. Indeed, these toxins behave more like chemical poisons.

Chemical weapons, such as nerve agents like sarin, blistering agents like nitrogen mustard, or poisons like cyanides, would probably be involved with small-scale or limited attacks also. While they could easily result in some deaths, they would probably be involved with focused attacks. Many antidotes are available for treatment of chemical agents. But to use antidotes effectively, rapid identification of the agent is critical.

The bottom line is that no panic is indicated, as most threats are low and limited. Most important, they are manageable. If we panic and disrupt our lifestyle, then the terrorists win regardless of the death toll achieved. The psychology of fear can be awesome. It is the bioterrorists' greatest weapon unless we deny them that weapon. In order to put things in better perspective, keep in mind that up to fifty thousand people die every year in the U.S. from influenza. About forty thousand die in automobile accidents annually. We lost a handful of people in the recent anthrax attacks, yet the media coverage of the latter is disproportionate to the former. We must all realize that most bioweapons are difficult to obtain, manufacture, or even process for large-scale use.

The threat of bioterrorism caused by an individual is real, but in most cases it is limited. A government-backed assault might be larger in scale, but it would be considered an act of war, thereby evoking a major retaliatory response from the U.S. government. Such an attack is not as likely, in my opinion, as a small-scale one. Our future needs are clear: We need to reverse the paradigm in the health-care system. Instead of saving money we need to increase public-health funding. We need to enrich and expand the resources in law enforcement. We need to increase research into better drugs and state-of-the-art vaccines, and blocking agents against various instruments of bioterror. Most of all, we must maintain our active state of vigilance! Remember: *the best weapons against bioterrorism are vigilance, education, and knowledge!* We have indeed begun that paradigm change when President Bush created the position of Director of Homeland Security and appointed Tom Ridge as its first director. As Tom Ridge so eloquently put it, "We must create a blueprint to win the wars of the future." He went on to say, "America will prevail!" I wholeheartedly agree.

GLOSSARY

aerobic: with oxygen

aerosol: via the air

afebrile: without fever

anaerobic: absence of oxygen

ARDS: acute respiratory distress syndrome

ascites: fluid in abdominal cavity

axillary: under the arms

BT: bioterrorist or bioterror

bubo: inflammatory swelling of a lymph gland

capsule: a thick outer covering

CDC: Centers for Disease Control

cervical: related to the neck or cervix region

cipro: Ciprofloxacin

crustular: crusts

cutaneous: related to or affecting the skin

cyanosis: bluish discoloration

dermal: of or relating to the dermis (skin)

desquamation: peeling of the skin

dyspnea: shortness of breath

ecchymosis: black-and-blue discoloration caused by the escape of blood into tissue

edema: swelling

edematous: swollen

ELISA: enzyme-linked-immunosorbent-assay

encephalitis: inflammation of the brain

endocarditis: inflammation of the lining of the heart

epidemiology: the study of the spread of an epidemic

equine: relating to a horse

eschar: scab

exanthem: rash or eruption of the skin

FA: fluorescent antibody

febrile: with fever

flaccid: limp

fluctuant: compressible

fomites: inanimate objects that may be contaminated with infectious organisms

gas gangrene: anaerobic destruction of tissue with gas production

GI: gastrointestinal

gram-negative: red-staining bacteria

gram-positive: blue-staining bacteria

hematemesis: vomiting of blood

hemolysis: red blood cell destruction

hemorrhagic: bleeding

HEPA: high-efficiency-particulate-air

hepatomegaly: enlargement of the liver

hypotension: low blood pressure

IM: intramuscular

index case: first case

inguinal: related to or in the area of the groin

jaundice: yellowish skin color

meningitis: inflammation of the membranes around the brain or spinal cord

micron: a millionth of a meter

morphology: study of the form and shape of an organism

muco-purulent: rich with mucus and pus

myalgia: muscle pain

myonecrosis: muscle destruction

nasal: related to the nose

necrosis: death of tissue

pandemic: affecting a large portion of a population

papule: small solid elevation of the skin (such as a pimple)

pasteurized: heated to 145°F for thirty minutes

PCR: polymerase chain reaction

penicillinase: enzyme that inactivates penicillin

phagocytosis: ingestion by white blood cells

photophobia: sensitivity to light

pneumonic: related to lungs

pulmonary edema: accumulation of fluid in the lungs

pustule: a small elevation of the skin filled with pus

rhinorrhea: runny nose

septicemia: blood infection

serum: liquid portion of blood or plasma

trachea: windpipe

viremia: viral infection of blood

virulence factor: factor that makes germs cause disease

REFERENCES

Bailey and Scott's Diagnostic Microbiology, 10th Edition. B. A. Forbes, D. F. Sahm, A. J. Weissfeld, editors. St. Louis: Mosby, Inc., 1998.

Biological and Chemical Terrorism: Strategic Plan for Preparedness and Response. Atlanta: MMWR, CDC, 2000.

Defense Against Toxin Weapons. D. R. Franc. Fort Detrick, MD: U.S. Army Medical Research Command, USAMRIID, 1997.

Food Microbiology. M. P. Doyle, L. R. Beuchat, T. J. Montville, editors. Washington, DC: ASM Press, 1997.

Jawetz, Melnick & Adelberg's Medical Microbiology. G. F. Brooks, J. S. Butel, S. A. Morse, editors, 21st edition, Appleton and Lange, Stamford, CT, 1998.

Manual of Clinical Microbiology, 6th Edition. P. R. Murray, E. J. Baron, M. A. Pfalles, F. C. Tenover, R. H. Youken, editors. Washington, DC: ASM Press, 1995.

Medical Management of Biological Casualties, 3rd Edition. Fort Detrick, MD: USAMRIID, 1998.

New York State Chem-Bio Handbook, James' Handbook. F. R. Sidell, W. C. Patrick, T. R. Dashiell. Alexandria, VA: James Information Groups, 2000.

The Secret Life of Germs. Philip M. Tierno, Jr., Ph.D. New York: Pocket Books, 2001.

Staying Healthy in a Risky Environment. A. C. Upton, E. Grober, L. Goldfrank, R. Nadig, A. Grieco, P. M. Tierno, Jr., editors. New York: Simon & Schuster, 1993.